U.S. CHEMICAL SAFETY AND HAZARD INVESTIGATION BOARD

INVESTIGATION REPORT

I0488871

VENT COLLECTION SYSTEM EXPLOSION

(1 Critically Injured, 18 Medical Evaluations, Community Evacuation)

TECHNIC INC.

CRANSTON, RHODE ISLAND

FEBRUARY 7, 2003

KEY ISSUES:

- INCOMPATIBLE CHEMICAL MIXING

- PROCESS SAFETY REVIEW

- MANAGEMENT OF CHANGE

- PREVENTIVE MAINTENANCE

- EMERGENCY PLANNING AND RESPONSE

REPORT NO. 2003-08-I-RI

AUGUST 2004

Contents

Figures

Tables

Acronyms and Abbreviations

ACGIH   American Conference of Governmental Industrial Hygienists

AgCN   Silver cyanide

$AgNO_3$ Silver nitrate

AIChE   American Institute of Chemical Engineers

AIHA   American Industrial Hygiene Association

ANSI   American National Standards Institute

ASTM   American Society for Testing and Materials

BOCA   Building Officials and Code Administrators, Inc.

CAA   Clean Air Act

CCPS   Center for Chemical Process Safety

CFD   Cranston Fire Department

CFR   Code of Federal Regulations

CSB   U.S. Chemical Safety and Hazard Investigation Board

CWRT   Center for Waste Reduction Technology

DOT   U.S. Department of Transportation

DSC   Differential scanning calorimetry

EPA   U.S. Environmental Protection Agency

EPCRA   Emergency Planning and Community Right-to-Know Act

°F   Degrees Fahrenheit

FEMA   Federal Emergency Management Agency

FM   Factory Mutual Insurance Company

GDC   General duty clause (OSHA)

ICC   International Code Council

IDLH   Immediately Dangerous to Life and Health

IPCS	International Program on Chemical Safety
ISO	International Standards Organization
LEPC	Local emergency planning committee
MOC	Management of change
NFPA	National Fire Protection Association
ORNL	Oak Ridge National Laboratory
OSHA	Occupational Safety and Health Administration
PSM	Process Safety Management (OSHA)
RI–DEM	Rhode Island Department of Environmental Management
RMP	Risk Management Program (EPA)
SARA	Superfund Amendments and Reauthorization Act
SCBA	Self-contained breathing apparatus
USFA	U.S. Fire Administration
VCS	Vent collection system

Executive Summary

On Friday, February 7, 2003, at approximately 9:30 am, an explosion ignited a fire inside a vent collection system (VCS) at Technic Inc., a plating chemicals manufacturing and research facility located in Cranston, Rhode Island. One employee was critically injured. Eighteen other employees were transported to Rhode Island Hospital for medical evaluation. The vent collection system was severely damaged, and plant operations were interrupted for several weeks.

In addition to evacuating plant employees, the Cranston Fire Department (CFD) ordered an evacuation of residents and businesses within 0.5 mile of the facility because of concern that cyanide salts and acids stored in the plant might combine to create toxic hydrogen cyanide gas.

According to witness testimony, the incident likely began when an employee—suspecting the plastic ventilation duct to be clogged—tapped on it with a small hammer. The resulting explosion severely injured his eyes and face. The U.S. Chemical Safety and Hazard Investigation Board (CSB) determined that the hammer blow likely initiated an explosive reaction of chemicals inside the duct.

The CSB investigation determined the following root causes:

- Technic did not conduct a process safety review to identify and evaluate the hazards associated with installing a vent collection system to handle the exhaust from multiple processes. Such a review would have likely included a process hazard analysis to identify and evaluate reactive chemical hazards. Competent reviewers would have evaluated and verified the design prior to and after installation.

- Technic did not identify and evaluate the hazards created by changes to facility processes and equipment (i.e., management of change). Significant changes made to the vent collection system were not envisioned in the original design and significantly deteriorated its performance. Such modifications should have been accompanied by a design and engineering review, a process hazard analysis, and before-and-after performance testing to ensure that the system was operating as intended.

In addition, the following contributed to the incident and its consequences:

- Neither Technic nor CFD adequately planned for this type of emergency.

- Technic did not have process and equipment integrity procedures or training to inspect or maintain the vent collection system.

CSB makes recommendations to Technic Inc., the National Fire Protection Association, the American Industrial Hygiene Association, and the American Conference of Governmental Industrial Hygienists.

1.0 Introduction

1.1 Background

A February 7, 2003, explosion and fire inside a vent collection system (VCS) at Technic Inc., in Cranston, Rhode Island, critically injured one employee, who suffered permanent eye damage and chemical burns to his face and upper body. Eighteen other employees were sent to the hospital for medical evaluations, and the fire department evacuated the surrounding community. Facility operations were interrupted for several weeks.

The explosion and fire were caused by a violent chemical reaction inside the vent collection system, which was likely initiated when the employee tapped on a duct with a small hammer. The building where the incident occurred housed several chemical processes that were connected to the ventilation system.

1.2 Investigative Process

The U.S. Chemical Safety and Hazard Investigation Board (CSB) arrived at the facility on the evening of February 7. CSB investigators examined physical evidence, conducted interviews, and reviewed relevant company documents. Samples of residue chemicals from interior surfaces of the vent collection system and associated equipment were collected and analyzed. CSB coordinated its investigation with the Occupational Safety and Health Administration (OSHA), U.S. Environmental Protection Agency (EPA), and Rhode Island State Fire Marshal.

The Rhode Island General Assembly passed a house resolution on February 26, 2003, requesting that CSB prepare a report of its findings regarding the Technic plant explosion. The report would contain the Board's recommendations and proposed policy changes to avoid further mishaps.

1.3   Vent Collection System Process Description

The vent collection system was installed in conjunction with a 1990 plant expansion. The system consisted of interconnected polyvinyl chloride duct segments, gradually decreasing in diameter, suspended approximately 20 feet above the floor. The purpose of the collection system was to transport vapors, gases, and mists emitted from various processes to a scrubber, in compliance with EPA air emission standards (Section 4.0).

The following processes connected to the vent collection system may have contributed to the incident (Figure 1):

- Five silver nitrate solution batch reactors located on mezzanine 1.

- The Kestner unit (a vessel used to manufacture silver nitrate solutions).

- Four potassium silver cyanide solution batch reactors located on mezzanine 2.

- More than 20 plating solution-mixing tanks/reactors located in the kettle room.

- Two lump removal and packaging stations for silver cyanide and potassium silver cyanide powders located in the weigh room.

- Eleven open surface tanks for gold/silver solutions in the metals reclaim areas.

With the addition of vent ducts from over 20 process vessels from 1992 to 2003, the demand on the vent collection system increased significantly from the time of the plant expansion in 1990. These additions were designed and installed without engineering analysis. See Appendix A for detailed descriptions of the chemical processes and emissions.

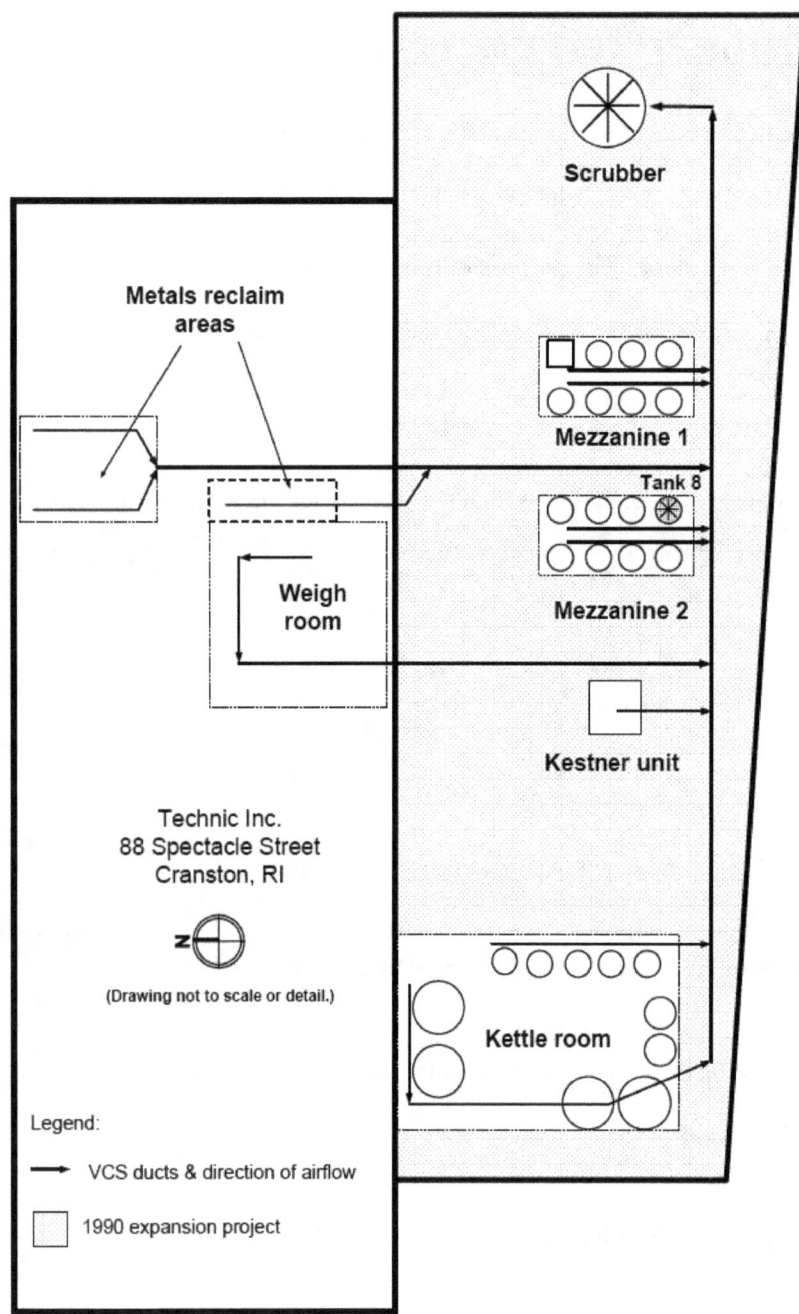

Figure 1. Vent collection system and connected processes.

1.4   Technic Inc.

Technic is a privately owned plating chemical and equipment manufacturing company founded in the 1940s. It is one of the largest volume producers of plating chemicals in the United States. Technic products are used for plating precious and nonprecious metals in the electronics and jewelry industries.

In addition to the Cranston facility, Technic has four other U.S. facilities—located in Anaheim, California; Plainview, New York; Clearwater, Florida; and Pawtucket, Rhode Island. The Anaheim and Plainview facilities conduct plating research and development, while the Pawtucket and Clearwater facilities make specialty equipment for the plating industry. Technic also has facilities in Europe and Asia, and employs over 500 people worldwide.

The Cranston facility employs 150 people, including several chemists. It supplies chemicals to over 1,000 customers in the United States and abroad. It is located in a mixed residential/industrial neighborhood alongside Rhode Island Route 10 (Figure 2). The facility includes two buildings, one of which houses the corporate offices and main research laboratories (1 Spectacle Street); the other houses the plating chemical manufacturing processes and related laboratories (88 Spectacle Street). The manufacturing building was expanded in 1990 to accommodate new production operations. The surrounding area included dwellings, businesses, and a recreational center 2 miles from the Technic facility.

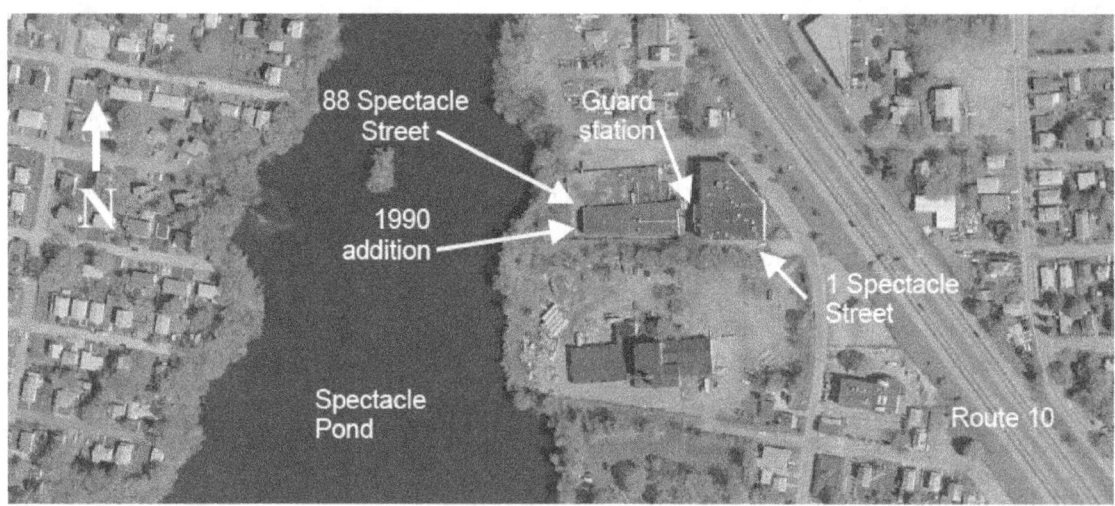

Figure 2. Aerial photo of Technic facility.

2.0   Incident Description

Manufacturing operations at Technic were significantly reduced on the day of the explosion. The production of silver cyanide salts had been temporarily suspended due to an employee absence. As a result, management had instructed employees to clean up their work areas.

The employee who was severely injured told CSB investigators that he was cleaning on mezzanine 2 when he heard a whistling sound coming from the exhaust duct connected to tank 8. He said it sounded as if the duct was blocked. After tapping on the duct with a small hammer, he thought he heard some loosened material fall into the tank. However, because the duct still sounded blocked, he tapped on it again. An instant later—at approximately 9:30 am—the duct exploded, knocking him to the floor.

Subsequent to the explosion, a fire involving accumulated combustible materials broke out inside the main vent header duct. The fire propagated through the main header duct to the branch ducts on mezzanine 1 and eventually consumed a plastic acid hood. According to the CFD report, the fire was knocked down by 11:34 am, and fire department personnel left the scene at 1:00 pm.

2.1   Employee Injuries

Five employees were working in the production area at the time of the explosion, but only the person who tapped the pipe was injured. Chemicals from inside the duct struck his face and upper body. After trying to flush his face and eyes at two inoperable eyewashes, he made his way to a sink-mounted eyewash in the adjacent building, where he was able to splash water onto his face until emergency responders arrived. He told CSB investigators that he was wearing wrap-around safety glasses at the time of the explosion and that he removed them sometime shortly afterward.

Emergency medical technicians from the Cranston Fire Department (CFD) arrived within 10 minutes of the explosion and began treating the victim. They cut off his shirt, washed his face and upper body with water, and used a saline-soaked cloth to remove chemicals and debris from his eyes, nostrils, and mouth. They did not flush his eyes with water because they were unable to determine from the employee, or other Technic personnel, exactly what chemicals he had been exposed to. Approximately 30 minutes after the first explosion, they decontaminated[1] the victim and placed him in an ambulance for transport to Rhode Island Hospital.

The ambulance crew irrigated his face and eyes for the duration of the transport. Although the hospital was only 4.3 miles away, the trip took 45 to 60 minutes due to snow-covered roads. After being decontaminated a second time at the hospital, the injured employee was treated for chemical burns to his face, neck, and eyes. He also suffered lacerations to his face, respiratory tract irritation, and a broken finger. He remained in the hospital for 14 days and was discharged with diminished sight in both eyes. His right eye required multiple surgical procedures and extended treatment over the next year.

2.2   Facility Damage

The effects of the explosion and fire were largely confined to the vent collection system, with minor damage to connected equipment (acid hood, vent hood, scrubber). The CFD cut away a small section of the roof above the process area to access the smoldering fire beneath the roof deck. Section 5.0 presents a detailed analysis of the explosion and fire damage.

---

[1] Decontamination involved stripping the employee's remaining clothes and washing him with soap and water.

3.0 Incident Response

3.1 Employee Evacuation

According to Technic employees working in the production building at the time of the incident, the fire was preceded by a series of loud explosions. Smoke started to fill the building, and the sprinkler alarm sounded. The administrative assistant called the CFD at approximately 9:34 am. Employees evacuated to the guard station (see Figure 2) for a headcount. Two people were missing. Management personnel, including one person with self-contained breathing apparatus (SCBA), entered the building to search for missing employees; they found no one and exited.

3.2 Firefighting

Within 10 minutes of notification, CFD arrived on scene to find that the building's sprinkler system was operating, and the fire was confined to the new addition. An incident command post was established at the corner of Spectacle and Russe Streets. CFD began fighting the fire from the exterior with hoses. Technic's president approached the operations officer to introduce himself and began to discuss the situation; however, he was turned away because the officer was busy and did not recognize him as someone with pertinent facility information.

The sprinkler system was operating when the firefighters arrived. However, it was turned off after some discussion among the firefighters, Technic management, and the Rhode Island Department of Environmental Management (RI–DEM) inspector because of concern that cyanide salt and acid might mix to form hydrogen cyanide, a toxic gas. The fire flared up again on mezzanine 1. CFD unrolled the hose from the truck parked in the street and put it in position at the overhead door on the east side of the storage shed on mezzanine 1. After testing the air for organic vapors, the CFD hazmat team entered the building and extinguished the smoldering fires inside the ventilation duct near the scrubber using portable extinguishers. The maintenance supervisor—wearing SCBA and accompanied by firefighters—then entered the building to shut down a process unit.

3.3 Decontamination of Exposed Employees

The CFD hazmat team set up a decontamination station as a precaution. Eighteen Technic employees who may have been exposed to cyanide or other potentially toxic chemicals were decontaminated and transported to Rhode Island Hospital for medical evaluation. At the hospital, they were decontaminated a second time, examined, and released; no one exhibited any symptoms of chemical exposure.

3.4 Community Evacuation

This incident occurred at mid-morning on a weekday, when many residents were away from their homes. However, because of concerns about smoke and potential cyanide gas moving in the general direction of the nearby community, the CFD incident commander directed the Cranston police to evacuate several businesses and residents downwind of the Technic plant. This decision was made approximately 15 minutes after CFD arrived on scene. At that time, the wind direction was from the northeast at 6 miles per hour. One dozen residents were bused to the Cranston Recreation Center, 2 miles away, in accordance with evacuation plans previously developed by the Region 8 Local Emergency Planning Committee (LEPC). The residents were sheltered for about 4 hours.

4.0   Vent Collection System

4.1   Design, Permitting, and Installation

In response to Clean Air Act (CAA) emissions standards, in 1990 Technic began a project to design the vent collection system to collect process exhausts and transport them to a scrubber. Using process descriptions and information supplied by Technic, a design consultant created a preliminary VCS design package to collect gases, mists, and vapors.[2] Although not accounted for in its design, the vent collection system also collected dusts. Duct size ranged from 36 inches in diameter at the entry to the scrubber to 4 inches at the process vessels.

By interviewing the design consultant and reviewing relevant ventilation design information, CSB investigators determined that the initial design did not provide capacity for processes that were added at a later date. Additions to the vent collection system after the original construction included the following:

- Ventilation ducts from the process reactors on mezzanine 1

- Plating tanks in the waste storage area

- Grinding and packaging operations in the weigh room

- Tanks in the kettle room

- Tanks in the silver oxide area.

In a cover memorandum accompanying the preliminary design package, dated December 21, 1991, the consultant asked Technic to clarify several process issues pertaining to hood design and airflow, and fire

---

[2] The minimum transport velocity required for gases, vapors, and mists is 2,000 feet per minute. The Technic VCS design package indicated that the main header duct was designed to this standard.

and explosion protection. According to the consultant, Technic did not respond; and the consultant had no further involvement with the project.

Using the preliminary design information, Technic applied to RI–DEM for a minor source construction permit on October 24, 1991, for the new vent collection system and scrubber. It was approved on January 16, 1992. The approval letter advised Technic to notify RI–DEM in writing before making any significant physical or operational changes to the system or connected process equipment.

Using information from the consultant's preliminary design package, workers from the Pawtucket facility fabricated and installed the vent collection system because they had experience constructing and installing ventilation ducts for plating systems. Once the ducts were installed, a licensed sprinkler contractor added a water sprinkler system inside the larger-diameter duct sections in accordance with the City of Cranston mechanical code.[3]

4.2   System Modifications

The minor source construction permit for the vent collection system included only the processes on mezzanine 2, the Kestner unit, the kettle room, the weigh room, and the metals reclaim areas. Technic had made a number of significant changes to the ventilation system since its installation in 1992. These modifications were not anticipated by the original design and construction permit[4] and diminished the system's performance capacity. Examples of significant changes include:

- Connecting seven process reactors and an acid hood on mezzanine 1.

---

[3] At the time the sprinklers were installed, the City of Cranston had adopted and was enforcing the 1987 edition of the Building Officials and Code Administrators, Inc. (BOCA), *National Mechanical Code.*

[4] In reviewing files, CSB found no evidence of notifications to RI–DEM for proposed changes to the ventilation system. When questioned on this point, Technic management personnel explained that they believed the changes did not trigger the major modification requirement of RI–DEM air pollution regulations. CSB believed that the modifications were major.

- Connecting dust-producing delumpers with simple filters and packaging equipment with an in-line baghouse in the weigh room.[5]

- Connecting additional process reactors and blend tanks in the kettle room.

- Connecting hoods for the three plating tanks in the metals reclaim areas.

- Removing the in-line baghouse for the delumpers in the weigh room in an attempt to increase the dust capture efficiency of the packaging equipment.

Technic maintenance personnel made these modifications in-house without consulting ventilation design or engineering professionals, and without referencing good ventilation practice guidelines. The result was a diminished transport velocity that allowed substances to settle out and accumulate inside the vent collection system.

Figure 3 depicts the vent collection system and its connected equipment as it was configured on the day of the explosion. This diagram is based on the original design drawings, with added observations and information obtained from CSB interviews.[6]

4.3   Loss Prevention Carrier Review

In April 2001, the Technic loss prevention insurance carrier recommended the installation of additional sprinkler heads inside all VCS ducts of 10-inch diameter or greater. This recommendation was referred to senior management but no action was taken. The 10-inch branch duct connected to the acid hood on mezzanine 1—which was destroyed in the fire—did not have sprinklers. In addition, the carrier had also asked that any future VCS changes be submitted to its field engineering group for review and comment.

---

[5] CSB was unable to determine the efficiency of these filtering devices or their capacity to prevent dusts from being deposited inside the vent collection system. The devices were added in 1992.

[6] Only the original design drawings were available to CSB; the drawings had not been updated to include 11 years of modifications.

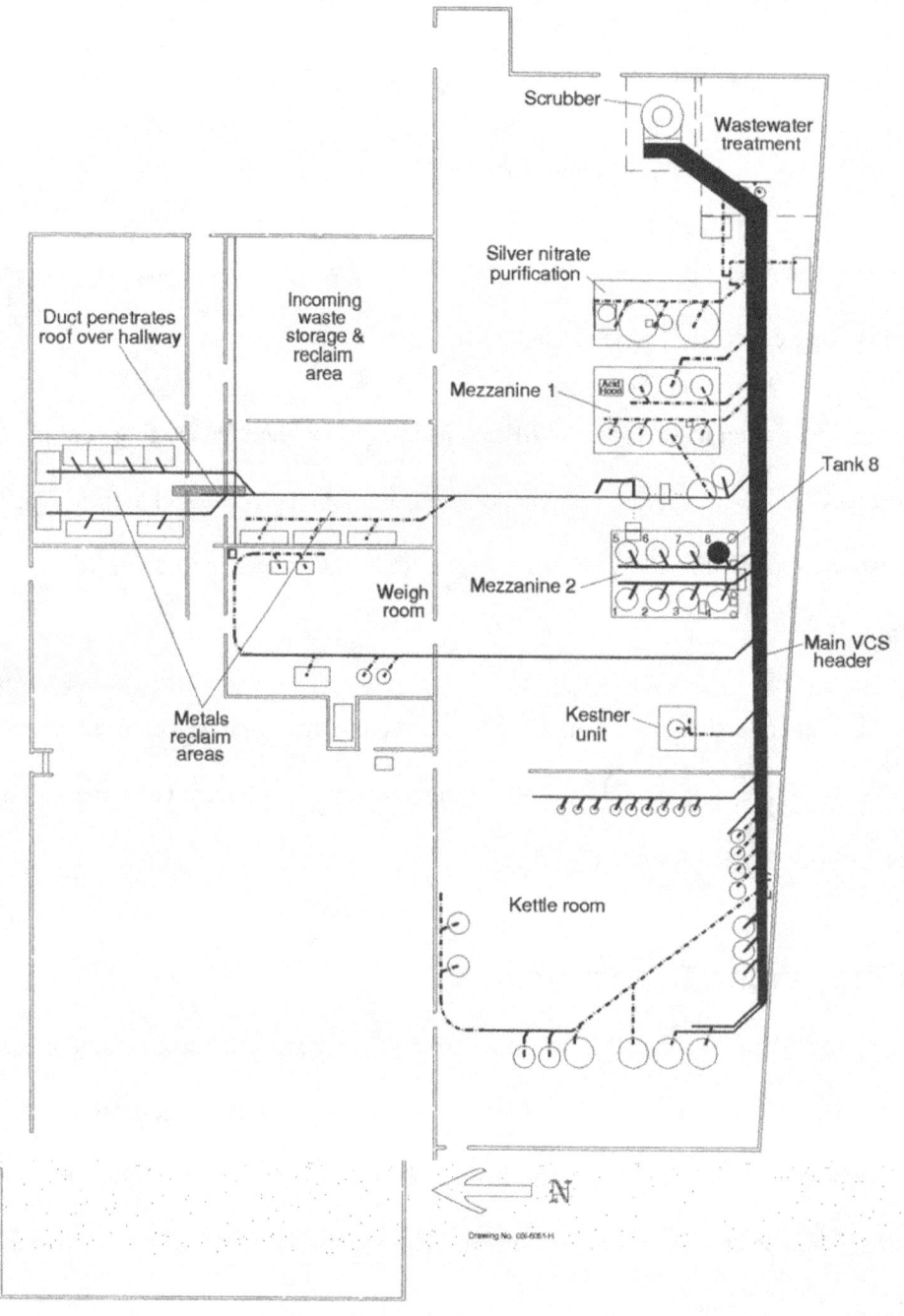

Figure 3. Vent collection system as of February 7, 2003
(modifications to original design are shown by dashed lines).

From April 2001 until the date of the explosion, additional kettles were added in the kettle room, and the baghouse in the weigh room was disconnected in an attempt to increase the duct capture efficiency of the packaging equipment. Discussions with the loss prevention carrier revealed that none of these modifications were submitted to the carrier for review.

5.0 Incident Analysis

CSB investigated the explosion and fire damage, and analyzed chemical reactions to determine the origin and causes of the explosion and fire.

5.1 Explosion and Fire Analysis

The explosion likely originated at the vertical exhaust duct connected to tank 8 on mezzanine 2 (Figures 1 and 3). This section of duct was shattered into small fragments that dispersed in all directions. Figures 4 through 8 show the extent of the explosion and fire damage to the vent collection system and its connected equipment. Table 1 outlines the sequence of events in the explosion and fire.

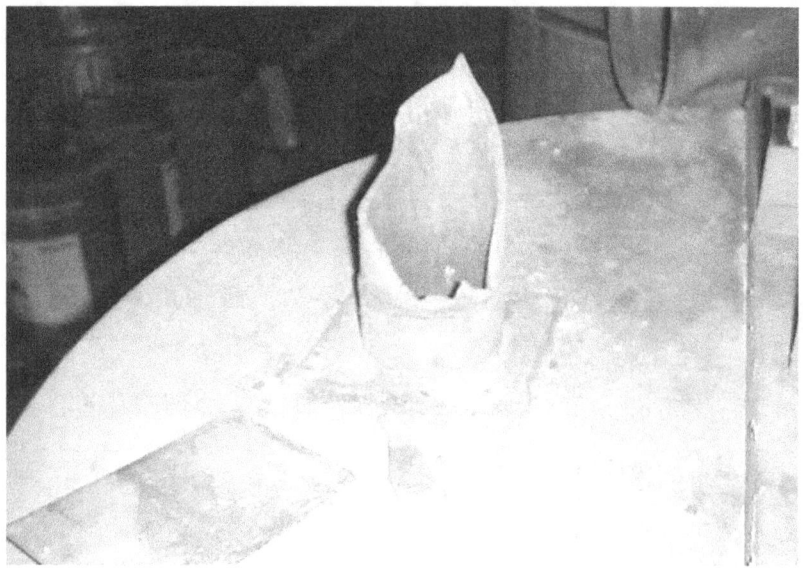

Figure 4. Sharp fracture of 4-inch exhaust duct above tank 8.

Figure 5. Large fracture of 10-inch branch duct above mezzanine 2 production area.

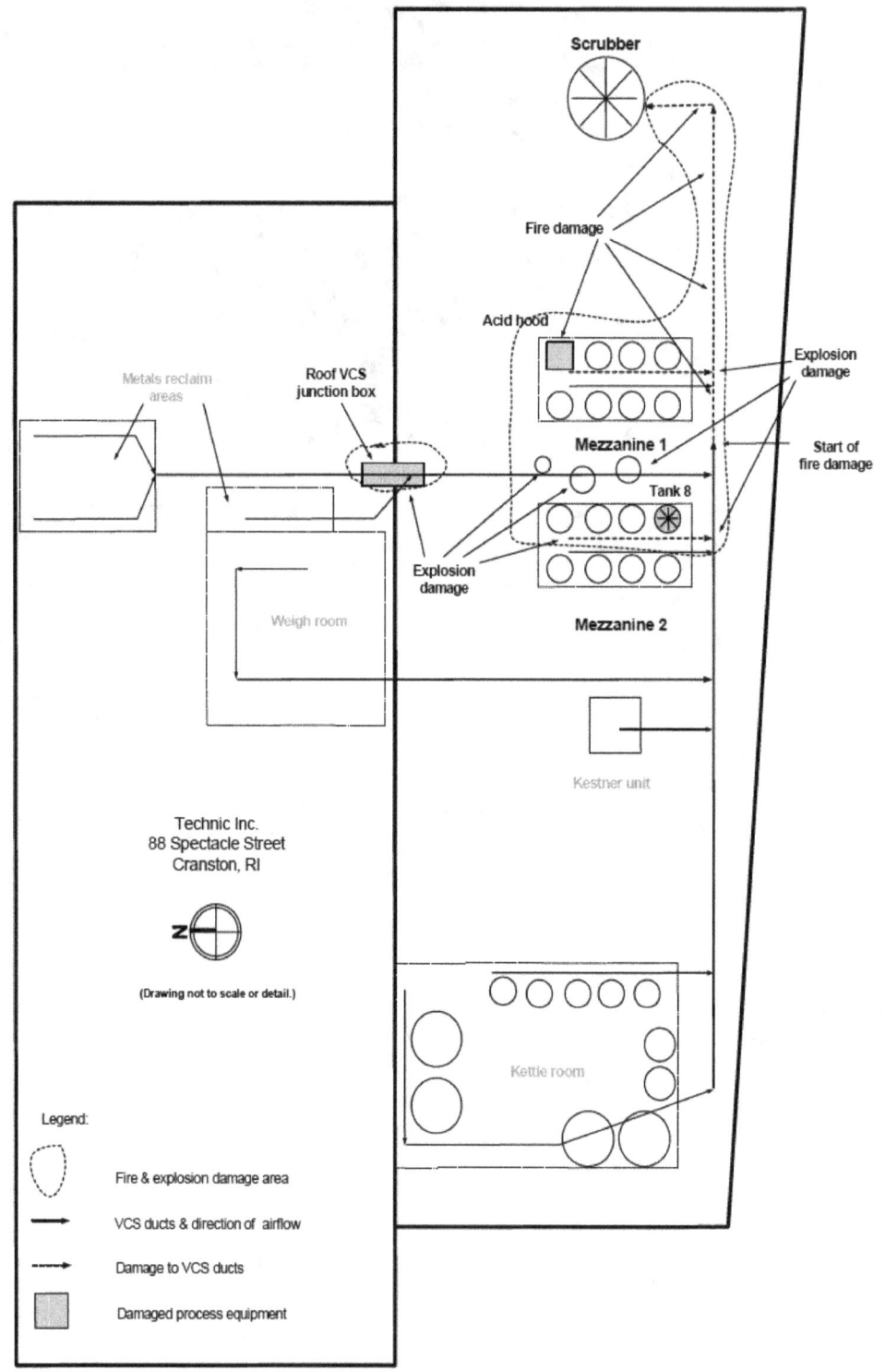

Figure 6. Depiction of explosion and fire damage, February 7, 2003.

Figure 7. Remains of acid hood on mezzanine 1.

Figure 8. Collapsed 36-inch-diameter main header at scrubber along east wall.

Table 1

VCS Explosion and Fire Damage Detail

Location	Damage	Discussion
Mezzanine 2	4-inch-dia vertical duct connected to tank 8 fractured and shattered (Figure 4)	Explosion originated inside exhaust duct, initiated by hammer blow
Mezzanine 2	10-inch-dia overhead horizontal branch duct connecting tanks 5, 6, and 7 (Figure 5) had a large crack at main header connection and sharp fractures where vertical ducts were attached	Pressure from explosion traveled against airflow
Mezzanine 2	4-inch vertical ducts connected to tanks 5, 6, and 7 blown off overhead horizontal branch duct	Damage limited to western overhead branch duct
Area between mezzanines 2 and 1	4-inch vertical duct connected to silver cyanide reactor blown off where it connected to horizontal branch header	Pressure from explosion traveled against airflow
Area between mezzanines 2 and 1	Hole blown out of 90-degree elbow connecting 4-inch vertical duct to silver cyanide centrifuge	Pressure from explosion traveled against airflow
On roof, large-diameter horizontal branch duct to metals reclaim areas	Roof junction box blown out on all sides	Pressure from explosion traveled against airflow
Mezzanine 1	Main header duct separated and collapsed (Figure 9)	Second explosion in main header duct; subsequent fire melted duct; deposited solid material provided fuel for fire
Mezzanine 1	Westernmost branch duct leading to tanks blown out at its connection to main header and in several other locations	Force generated by second explosion blew out branch junctions upstream, against airflow
Mezzanine 1	Portions of easternmost branch duct blown out at tank junctions (a)	Force generated by second explosion blew out branch junctions upstream, against airflow
Mezzanine 1	Plastic acid hood and its contents totally destroyed by fire (Figure 7)	Fire traveled to acid hood, against airflow, and destroyed plastic hood and its combustible contents
Above northern-most end of mezzanine 1	Firefighters opened this section of roof to extinguish fire	Radiant heat from burning acid hood charred building roof
Main header from mezzanine 1 to scrubber	50 feet of duct melted and collapsed (Figure 8); hazmat team opened section of duct to extinguish smoldering fire	Duct contained large amount of deposited combustible material that continued to smolder and flare

(a) CSB investigators were unable to examine a large portion of the branch duct connecting the acid hood to the main header because it was missing.

The initial explosion likely ignited fuel (e.g., dusts or accumulated residue) inside the main header. The resulting flame moved downstream toward the scrubber, causing at least one additional explosion. The presence of shiny, blistered char inside the duct indicates that the fire burned intensely for an extended time (Figure 9).[7]

Figure 9. Blistered char inside 36-inch-diameter main header adjacent to mezzanine 1.

The fire destroyed approximate 50 feet of the main header. No explosion or fire damage was identified upstream of mezzanine 2.

5.2 Chemical Reaction Analysis

CSB examined all of the process areas connected to the vent collection system and the chemicals used in each area. Through years of accumulation of solids (Figure 10) and mixing, many materials could have resulted in violent reactions such as the one that caused this incident. CSB evaluated only a sample of

---

[7] Char is a carbonaceous material that has been burned and has a blackened appearance. When char is shiny with interlaced cracks, it is commonly referred to as "alligator char."

those materials that are used in large quantities and might have accumulated in the collection system—silver nitrate, silver cyanide, sodium hydrosulfite, ammonium citrate, carbon (representing the many organic chemicals in the plant), and ammonia.

Figure 10. Solid deposits in main header duct.

The injured employee was the only eyewitness who provided CSB with information pertaining to the initiating event—namely, that he tapped on the duct with a small hammer, and the explosion immediately followed. The CSB investigation focused on substances capable of reacting violently following the application of a relatively small amount of energy.[8]

In collaboration with a reactive chemical expert hired by CSB, OSHA analyzed more than 70 samples from locations inside the vent collection system and affected areas, and from the injured employee's shirt. Two scenarios emerged that could account for the initial explosion:

---

[8] Tapping of the hammer would produce significantly less energy than a lighted cigarette, electrical spark, or open flame.

- A violent reaction of a shock-sensitive material

- A violent reaction between silver nitrate and another compound.

Although neither scenario can be conclusively identified as the initiating event, CSB asserts that the factors were present to support either reaction. Several other reaction scenarios were considered and discarded.

The *International Chemical Safety Card*[9] for silver nitrate states that it is a "strong oxidant" and "reacts violently with combustible and reducing materials," forming "shock-sensitive compounds" (International Program on Chemical Safety [IPCS], 1993). Silver nitrate is routinely carried into the vent collection system together with combustibles and reducing materials.

5.2.1   Shock-Sensitive Explosive Silver Compounds

This scenario examines the potential for the presence of shock-sensitive explosive silver compounds inside the vent collection system. These compounds are highly unstable materials—even a small amount of energy (e.g., the force created by a hammer blow) can cause detonation.

- ***Compounds formed from silver nitrate and ammonia***—"Many unstable silver compounds are known whose decomposition can occur spontaneously and explosively." Renner (1993) lists two compounds—silver azide and silver amide—that could have been formed from silver nitrate and ammonia.

    - Silver nitrate is produced daily on mezzanine 1 and in the Kestner unit. It is transferred to tank 8 before being used in potassium silver cyanide production.

---

[9] International chemical safety cards are designed to supply workers with information on the properties of chemicals. They are authoritative documents written and extensively peer reviewed by internationally recognized safety experts.

- Ammonia is emitted from two plant processes. It is produced in the reactors on mezzanine 2 and during treatment of potassium silver cyanide solutions in the metals reclaim areas.

Although a trace amount of silver azide was identified in one sample taken from inside the vent collection system, none was found in any of the ducts involved in the explosion or fire. Therefore, it is uncertain whether silver azide played a role in the initial explosion. The force of the explosion, subsequent pressure waves, and heat from the fire would likely have initiated and consumed any trace amounts of the substance.

- *Metallic silver and silver nitrate*—Shock-sensitive explosive silver compounds can form interlinked mixtures of metallic silver and silver nitrate (Ennis and Shanley, 1991). Although elemental silver was found in several samples—and silver nitrate was routinely introduced into the vent collection system—analytical results did not identify any shock-sensitive metallic silver or silver nitrate compounds.

5.2.2   Silver Nitrate Reacting With Another Compound

There is some evidence to support the possibility that silver nitrate reacted directly with another compound:

- Silver nitrate is a known oxidizer and will react with organic compounds (National Fire Protection Association [NFPA], 1998).

- Silver nitrate is present in the manufacturing process and likely in the vent collection system.

- Various organic compounds have been introduced into the vent collection system over the years—many of which were in the form of dusts.

- Ammonium citrate and sodium hydrosulfite—compounds used in the manufacturing process in significant quantities—were tested with silver nitrate using differential scanning and accelerating rate calorimeters. In each case, it was determined that a hammer blow could initiate the reaction and produce the damage observed in this incident. (See Appendix B.)

- Elemental silver was found in the vent collection system, though it is not used in the manufacturing process in a form likely to be deposited. Elemental silver is produced from the reaction of silver nitrate with another organic compound.

6.0   Analysis of Safety Management Systems

Process safety management systems help ensure that facilities are designed, maintained, and operated in a safe manner (Center for Chemical Process Safety [CCPS], 1989). Although neither the OSHA Process Safety Management (PSM) Standard nor the EPA Risk Management Program (RMP) was applicable to the Cranston Technic facility, management systems are necessary to ensure that process safety procedures are followed.

In comparing Technic management systems with CCPS and other industry guidance and standards, CSB concluded that safety management system failures were the underlying cause of the explosion and fire at Technic. CSB reviewed the following safety management systems:

- Process safety review for projects

- Management of change (MOC)

- Process and equipment integrity

- Emergency planning and response.

6.1   Process Safety Review for Projects

Good engineering practices include the use of appropriate codes and standards. As part of the engineering process, formal documented process safety review procedures are necessary to ensure an effective safety management program and the application of good engineering procedures. Competent engineering and safety personnel must be involved to ensure that these procedures explicitly cover the identification and mitigation of hazards. Although the Cranston facility had the technical and operational personnel to conduct a process safety review, there was no system in place that required the review.

An initial hazard review should be conducted during the early stages of a project to identify incompatible materials and the need to separate them. CSB found no evidence of such a review for the vent collection system, though the company asserts that there were several safety discussions during the VCS project.

Because of the failure to follow a formalized process safety review procedure, Technic installed a vent collection system that:

- Did not meet regulatory or industry good practice guidelines

- Was not capable of being modified without degrading performance

- Allowed incompatible chemicals to accumulate in the duct and mix.

- Did not address the concerns of the insurance carrier and the ventilation design contractor.

6.2   Management of Change

MOC is a system of managing process system changes (e.g., chemistry, equipment, procedures, or organization) to prevent catastrophic consequences. MOC allows for the identification of hazards so that they can be limited or mitigated. Even a minor process change may trigger a formal process hazard review to identify and evaluate its effects. Process hazards created by such changes must be documented and mitigated through engineering or work-practice controls.

Technic did not conduct process hazard reviews in conjunction with process changes. Significant modifications made to the vent collection system over its 11 years of operation severely degraded transport velocity and introduced a host of chemicals—including dusts—that were not evaluated for compatibility. A process hazard review would likely have identified silver nitrate and organic compounds as incompatible, signaling the need for separate ventilation control systems.

For example, the vent collection system was not designed with enough velocity to transport dusts generated in the weigh room to the scrubber; it was designed only for gases, vapors, and mists. In-house maintenance personnel installed filtering devices near the dust-generating equipment.[10] However, the large amount of accumulated debris in the main VCS header—downstream of the weigh room branch duct—indicates that the filtering devices were either improperly sized or inadequately maintained. Process modifications and duct additions to the ventilation system over the years further deteriorated transport velocity—which was inadequate for dusts to begin with. The large amounts of dust that accumulated were available to react with other substances passing through the system and also provided fuel for the February 7 fire.

In December 2002, OSHA cited Technic for employee overexposure to silver dusts in the weigh room. After receiving the citation, Technic attempted to eliminate these exposures by removing the baghouse that captured dusts generated during packaging and weighing. Technic management told CSB that the baghouse created a drag on the system; by removing it, they expected the capture velocity to improve enough to reduce employee exposure to within acceptable limits.

This change did not trigger a process hazard review by Technic management. Silver levels in the silver recovery area—which handles effluent from the scrubber—immediately spiked, signaling a significant increase in silver-containing dusts being introduced into the vent collection system.[11] This result was likely due to removal of the filters.

---

[10] A baghouse was installed 10 to 15 feet downstream of the weigh room, and household heating system filters were installed approximately 5 feet downstream of the delumpers.

[11] There is some controversy as to the relationship between disconnecting the baghouse and the spike in silver levels. The Technic waste treatment technician claims that he noticed the spike immediately, which was confirmed by Technic management. A review of the scrubber water treatment logs, however—though indicating a noticeable increase seemingly associated with disconnect of the baghouse—also reveals that in several periods while the baghouse was connected and in use, there were also spikes in silver effluent. It is worth noting that these spikes may have been due to inconsistent baghouse maintenance or other operator-related factors.

6.3   Process and Equipment Integrity

The objective of a process and equipment integrity program is to identify process and equipment defects that could result in a catastrophic failure. A comprehensive program will ensure the safe operation of processes and equipment throughout their life cycles. Technic had some elements of a process and equipment integrity program on other equipment, but there was no formal, scheduled, and documented inspection and preventive maintenance program for the vent collection system.

Because the actual performance of the vent collection system was not measured, the facility had no baseline—other than theoretical values in the original design drawings—against which to measure VCS performance. The collection system had no inspection ports or cleanout doors, and it had not been cleaned since 1996. These design features—combined with a routine inspection and maintenance program—would have facilitated the detection and elimination of accumulated material that CSB observed and believes likely contributed to the initial explosion and fueled the fire. In summary, Technic had no established inspection, cleaning, or maintenance program for the vent collection system.

6.4   Emergency Planning and Response

Facilities that handle highly hazardous materials should be designed with multiple layers of safety protection to ensure the safety of workers and the surrounding community (CCPS, 1995; Figure 11). On the day of the incident, the inner layers of protection at Technic failed (i.e., design and engineering, and process and safety controls), causing the facility to rely on mitigation and emergency response. Although the inner layers of protection offer more prevention than the outer layers, all layers are necessary.

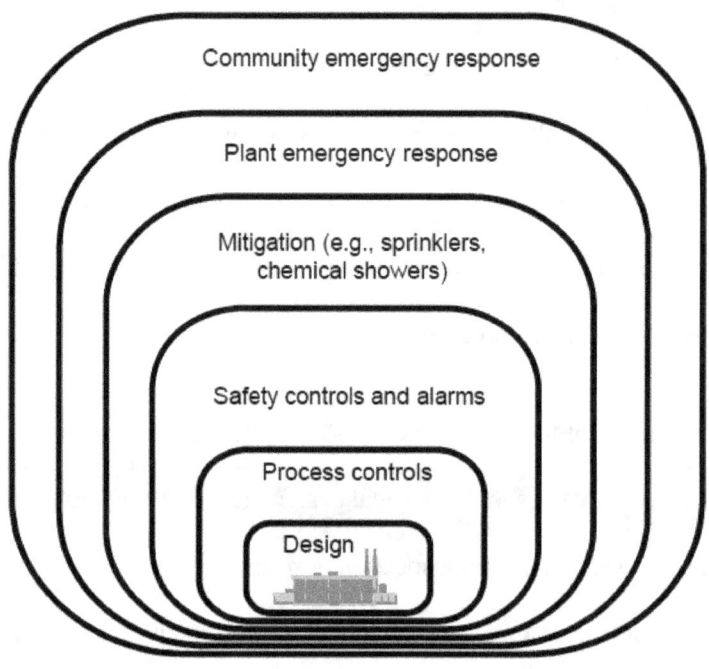

Figure 11. Layers of protection in a modern chemical plant.

6.4.1   Facility Emergency Planning and Response

6.4.1.1   Emergency Plan

Emergency plans are mandatory under both OSHA and EPA[12] regulations; generally accepted industry

guidance is available from sources such as CCPS (1995). Prior to the incident, Technic had provided the

municipal fire department (CFD) with an emergency plan for its facility. However, as the incident

revealed, the plan did not stand well on its own and did not meet minimum regulatory requirements.

Prior coordination and rehearsal by Technic and CFD would have revealed the plan's shortcomings and

likely improved response.

---

[12] The OSHA standard for emergency action plans is codified at 29 CFR 1910.38(a)(2). EPA requirements are
codified in the Emergency Planning and Community Right-to-Know Act of 1986 (EPCRA), 40 CFR 350–372.

For instance, the emergency plan did not describe the exact quantities and locations of cyanides and acids, or the provisions taken by Technic to safeguard them. CFD may have responded less aggressively to the fire than it otherwise would have because the firefighters feared that hydrogen cyanide—an extremely toxic gas—could be generated if sprinkler water accumulating on the floor somehow allowed the mixing of cyanide salt and acids. The sprinkler system was perhaps shut off prematurely. If the plan had specified that hazardous materials were invulnerable to firefighting because they were separated or contained (as was the case at Technic)—or if the plan had been rehearsed or coordinated in advance—CFD may have taken more aggressive action sooner.

Likewise, the emergency plan did not outline important roles and responsibilities—such as the Technic point of contact, the incident commander, the person responsible for notifications, or the person responsible for corrective action (e.g., shutting down critical processes and ensuring that all personnel are counted for). As discussed in Section 3.2, the CFD operations officer did not know who was in charge at Technic. Planning, coordination, and training can help minimize stress and confusion by removing uncertainty and improving confidence.

6.4.1.2   Search and Rescue

The employee evacuation was disorganized and confusing, which resulted in unnecessary, hazardous actions by Technic employees. A preliminary head count revealed that two employees were missing. A supervisor donned SCBA and reentered the process building to search for them. This was in violation of established OSHA regulations pertaining to fire prevention and respiratory protection, as discussed below.

OSHA considers search and rescue conducted during a fire to be advanced firefighting, which is addressed in the Fire Brigade standard (29 CFR 1910.156). Technic did not have a fire brigade, or the requisite equipment or training to carry out its functions. Technic personnel were trained only in the use of portable fire extinguishers to put out incident stage fires (29 CRF 1910.157). In addition, the procedures for the use of SCBA in firefighting situations—outlined in the OSHA Respiratory Protection standard (29 CFR 1910.134 (g) (3))—were not followed. These procedures require at least four persons (two to enter and two as standby) to be equipped with SCBA when entry is made into an unknown atmosphere (e.g., immediately dangerous to life or health [IDLH]) during firefighting. Periodic evacuation drills would have likely resulted in employees being accounted for and thus precluded unnecessary search and rescue efforts.

6.4.2   Community Emergency Planning and Response

According to EPA, no one was exposed to hazardous substances. CFD initiated community evacuation at the first sign of any hazards, and all residents in the zone of danger were successfully evacuated to a point of safety. CFD prepared an annual emergency plan for Technic, which was carried on its fire trucks. The community emergency response was consistent with established guidelines (Federal Emergency Management Agency [FEMA], 2002).

The Cranston Fire Chief is also chairman of the Rhode Island District 8 LEPC. The purpose of the LEPC is to plan and coordinate community protection against catastrophic chemical releases based on EPCRA requirements (see Section 7.1.5). The fire chief told CSB investigators that, based on Technic's chemical inventory submissions, it is among the top 10 facilities of concern within District 8.

Because of other priorities such as homeland security, the LEPC had not held public meetings in a year, was behind in its EPCRA-required community emergency response plan, and had never conducted an emergency response exercise with Technic. Specifically, the community plan lacked facility-specific information for an adequate emergency response, such as location of chemicals, hazards of mixing chemicals, and details on the containment and segregation of chemicals. The fire chief told CSB investigators that these deficiencies were due to a lack of funding, personnel, and computer equipment— and to conflicting national priorities for terrorism planning among some LEPC members (e.g., police, fire department, hospitals). The involvement of the LEPC and CFD in the planning process, as required, would have improved emergency response.

7.0   Regulatory and Industry Standards Analysis

CSB reviewed environmental, safety, and health guidance applicable at the time (1992) Technic installed the vent collection system. The review included government regulations, local building and mechanical codes, and industry consensus standards (guidelines) relevant to the explosion and fire. CSB also looked at currently applicable guidance to determine if revisions are warranted. Only the guidance applicable to the Technic incident is discussed below.

7.1   Environmental, Safety, and Health Regulations

7.1.1   OSHA Process Safety Management

The OSHA PSM Standard (29 CFR 1910.119) is a systematic approach to process safety and the prevention of catastrophic incidents. It requires adherence to 14 elements of good safety management for processes containing specific chemicals or flammables if they are present in quantities greater than 10,000 pounds. The PSM Standard does not apply to Technic because listed chemicals are not present at the required threshold quantities.

7.1.2   EPA Risk Management Program

The EPA RMP is a product of the CAA Amendments of 1990. In many ways, RMP mirrors the PSM Standard. However, RMP focuses on protecting the public and the environment from offsite consequences, while PSM addresses consequences for workers. RMP does not apply to Technic because listed chemicals are not present at the required threshold quantities.

7.1.3   OSHA Ventilation Standards

Although OSHA has no general safety or health regulations that apply to vent collection systems, it does have specific ventilation regulations for dipping and coating operations.[13] These regulations apply to the Technic metals reclaim areas, which are connected to the ventilation system. They require separate exhaust systems to prevent mixing of incompatible materials, as well as periodic inspection and maintenance.

For other types of ventilation systems, OSHA relies on the general duty clause (GDC) to enforce safe use. This clause—found in Section 5(a)(1) of the Occupational Safety and Health Act, promulgated in 1970— requires employers to eliminate recognized hazards in their workplaces, even when there is no specific OSHA rule or standard. According to OSHA, employers must take whatever abatement actions are feasible to eliminate hazards.

For the February 7 explosion and fire, OSHA cited Technic:[14]

- For its actions that created fire and explosion hazards within the vent collection system, under the GDC.

- For not providing independent exhaust systems for tanks in the silver recovery area, under the dipping and coatings standard.

7.1.4   Clean Air Act

Under authority of the CAA of 1970, RI–DEM administers a new source permitting program for air pollution sources like the vent collection system. This State program requires applicants to apply for a permit if they create a new source that generates any one of the EPA-listed hazardous air pollutants, or if

---

[13] 29 CFR 1910.123, Dipping and Coating Operations: Coverage and Definitions; and 29 CFR 1910.124, General Requirements for Dipping and Coating Operations.

[14] Citations were issued on July 28, 2003, and settled on August 11, 2003.

they intend to significantly modify an existing pollution source. The program focuses on compliance with air emissions regulations and does not address process safety hazards from incompatible chemical mixing.

RI–DEM conducted a technical review of the design drawings—but only from the perspective of emissions reduction. RI–DEM also conducted a post-installation site visit, but did not attempt to identify process changes or minor variations between design and actual installation.

7.1.5   Emergency Planning and Community Right-to-Know Act

The Superfund Amendments and Reauthorization Act (SARA), Title III, was enacted in 1986. EPCRA codifies EPA requirements—which call for each state governor to set up a State Emergency Response Commission, which establishes LEPCs.

LEPCs are charged with developing community emergency response plans for incidents involving hazardous materials. They also routinely coordinate chemical release incident exercises and facility visits so that emergency response team members become familiar with the unique challenges of each facility. These activities help pinpoint plan deficiencies and promote confidence among team members and plant management.

As noted in Section 6.4.2, the Rhode Island Region 8 LEPC has been delayed in implementing the requirements of its community emergency response plan.

7.2 Mechanical Code

A mechanical code requires engineers and design professionals to incorporate safety features into the design, engineering, and construction of mechanical systems installed inside buildings. The State of Rhode Island had a mechanical code in effect when the Technic consultant designed the vent collection system, but the code did not prohibit incompatible reactive chemical mixing or require designers to ascertain whether such mixing might occur.[15] The City of Cranston building department—the local agency that reviews and enforces the code—reviewed the Technic VCS design, but relied heavily on the permit review conducted by RI–DEM. Like RI–DEM, the city did not assess chemical compatibility.

In 2002, the State of Rhode Island adopted a mechanical code that requires incompatible materials to be exhausted through separate exhaust systems.[16] Although this requirement provides much needed guidance to designers and engineers, and to building department officials and building inspectors who review VCS designs and installations, it has not been applied retroactively.

7.3   Consensus Standards

Consensus standards are voluntary compliance guidelines created by balanced working groups of interested and diverse experts and stakeholders. NFPA, the American National Standards Institute (ANSI), and the International Standards Organization (ISO) are examples of respected standard-making organizations. Although consensus standards are generally nonmandatory, they are often referred to (or adopted) in regulations and codes, and are used extensively throughout industry.

Rhode Island has adopted the Uniform Fire Code (NFPA 1), which incorporates NFPA 91 and NFPA 654 by reference. There are currently four widely used consensus standards pertaining to vent collection systems:

- NFPA 91, *Standard for Exhaust Systems for Air Conveying of Vapors, Gases, Mists, and Noncombustible Particulate Solids*: NFPA 91 applies to all exhaust systems—except those designed for combustible particulates—and is intended to provide life and property protection against fires and explosions. This standard specifically prohibits incompatible materials from

---

[15] As noted in Section 4.1 of this report, Cranston used the BOCA *National Mechanical Code*, 1987 edition.

[16] The Rhode Island Building Code Standards Committee adopted the International Code Council (ICC) *International Mechanical Code* on September 1, 2002. Section 510.5 addresses incompatible materials.

being mixed. Other requirements of the standard are access doors in horizontal ducts every 12 feet; automatic fire extinguishing systems inside ducts; and testing, inspection, and maintenance of these systems at regular intervals (NFPA, 1999). NFPA 91 is deficient in that it does not reference appropriate methods for such evaluation, such as ASTM E-2012-0.

- NFPA 654, *Prevention of Fire and Dust Explosions From the Manufacturing, Processing, and Handling of Combustible Particulate Solids*: NFPA 654 supplements NFPA 91 by specifically addressing hazards associated with systems that handle combustible dusts. It requires conformance with all the provisions of NFPA 91, in addition to preventive measures to eliminate the accumulation of materials in ducts and the propagation of explosions through the duct system (NFPA, 2000).

- ANSI Z9.2, *Design and Operation of Local Exhaust Ventilation Systems*: ANSI Z9.2 applies to vent collection systems used to control employee exposures to harmful substances. It requires that separate exhaust systems be used when incompatible material mixing may result in health or explosion hazards, or destructive corrosion (ANSI, 1991). ANSI Z9.2 is deficient in that it does not reference appropriate methods for such evaluation, such as ASTM E-2012-0.

- *Industrial Ventilation: A Manual of Recommended Practice*: This American Conference of Governmental Industrial Hygienists (ACGIH) manual is widely used to design and evaluate new and existing industrial ventilation systems. Although it instructs designers to take preliminary steps to ascertain toxicity, and physical and chemical characteristics, it does not specifically call for consideration of chemical compatibility (ACGIH, 2001).

The mechanical code and some consensus standards underscore the need to identify incompatible materials. However, none of these resources outlines a method to conduct such analyses. The CSB hazard study *Improving Reactive Hazard Management* (USCSB, 2002) suggests using ASTM E 2012-00,

Standard Guide for the Preparation of a Binary Chemical Compatibility Chart (American Society for Testing and Materials [ASTM], 2003).

The ASTM guide provides a systematic method for determining whether two chemicals will react with one another under a stated scenario. If used during the planning and design phases, or in conjunction with a process hazard review, this particular binary evaluation system will often identify incompatibility hazards that would otherwise be missed. Although the method applies only to two materials, it can be adapted to include multiple materials. Moreover, in addition to providing a structure for incompatibility analysis, it documents the steps necessary to identify and evaluate hazards. It also can be used to demonstrate to regulatory agencies that such an analysis has been conducted.

7.4   Loss Prevention Guidance

The Technic loss prevention insurance carrier published a property loss prevention data sheet that lists recommendations for industrial exhaust systems (FM Global, 2000). It addresses the following:

- Transport velocities

- Materials of construction

- Sprinkler protection

- Incompatible materials

- Inspection and maintenance.

Although this data sheet was not available during design and construction of the vent collection system, it was presented to Technic prior to the incident (April 2001) in support of a recommendation to install additional sprinkler heads in the VCS ducts. Receipt of the data sheet should have signaled the need to evaluate fire and explosion hazards by conducting a process safety review. Technic took no action, explaining to CSB that it believed the system was performing adequately. Emissions were being reduced to permit levels, and there had been no prior incidents that signaled potential safety hazards.

8.0 Root and Contributing Causes

8.1 Root Causes

1. **Technic did not conduct a process safety review as a part of the engineering process to identify and evaluate the hazards associated with installing a vent collection system to handle the exhausts from multiple processes.**

   - The ventilation consultant hired to design the system posed several questions to Technic regarding the materials it intended to introduce into the system. Technic did not respond to these questions and did not use the consultant during the remaining phases of the project.

   - Despite having chemists on staff, Technic did not take advantage of their expertise in assessing process chemical compatibility.

   - Such a review would have likely included a process hazard analysis to identify and evaluate reactive chemical hazards. It would have also involved competent reviewers to evaluate and verify the design prior to and after installation.

2. **Technic did not identify and evaluate the hazards created by changes to facility processes and equipment (i.e., management of change).**

   - Technic maintenance personnel modified the vent collection system without applying good engineering practices. Significant changes were made to the vent collection system that were not envisioned in the original design and deteriorated its performance. Such modifications were not accompanied by a design and engineering review, a process hazard analysis, or before-and-after performance testing to ensure that the system was operating as intended. The relative effects of these modifications were not evaluated against baseline performance measurements.

- Technic did not take advantage of the expertise and engineering review services of its loss prevention insurer.

8.2   Contributing Causes

1. **Technic did not have process and equipment integrity procedures or training to inspect or maintain the vent collection system. The system was built without cleanout doors and was not capable of easily being inspected. It had not been cleaned since 1996.**

2. **Neither Technic nor the Cranston Fire Department (CFD) adequately planned for this type of emergency.**

   - The emergency plan Technic provided to CFD did not include key facility information, and CFD was not familiar with the facility or its managers. This led to confusion and uncertainty during the response and likely accounted for the severity of the fire damage.

   - Neither Technic nor CFD had ever conducted a chemical release exercise. Technic was unable to account for all its employees, which led a supervisor to take risky actions to reenter the building.

9.0   Recommendations

Technic Inc.

1.   As a part of the engineering process, implement formal process safety review procedures for projects involving chemical processes—including the vent collection system. Incorporate a process hazard analysis, reactive chemical hazard evaluation, and design evaluation consistent with applicable codes and standards. (2003-08-I-RI-R1)

2.   Implement a management-of-change program and ensure that reviews are conducted for any proposed changes to the vent collection system and its connected processes. (2003-08-I-RI-R2)

3.   Implement a preventive maintenance program for the vent collection system that includes regular inspections training and troubleshooting. (2003-08-I-RI-R3)

4.   Work with the Cranston Fire Department to improve the facility's emergency response plan, including emergency response procedures and interface with the surrounding community. Submit the plan to the fire department for review. (2003-08-I-RI-R4)

National Fire Protection Association (NFPA)

Revise the appendix of NFPA 91, *Standard for Exhaust Systems for Air Conveying of Vapors, Gases, Mists, and Noncombustible Particulate Solids*, emphasizing the need to evaluate potential incompatibilities when dusts, fumes, or vapors are intermixed in vent collection systems to ensure that they do not result in fire or explosion hazards, or destructive corrosion. Reference appropriate methods for such evaluations, such as the ASTM E 2012-00 standard. (2003-08-I-RI-R5)

American Industrial Hygiene Association (AIHA)

Revise the appendix of ANSI Z9.2, *American National Standard Fundamentals Governing the Design and Operation of Local Exhaust Systems*, emphasizing the need to evaluate potential incompatibilities when dusts, fumes, or vapors are intermixed in local exhaust ventilation systems with common headers to ensure that they do not result in fire or explosion hazards, or destructive corrosion. Reference appropriate methods for such evaluations, such as the ASTM E 2012-00 standard. (2003-08-I-RI-R6)

American Conference of Governmental Industrial Hygienists (ACGIH)

Update the preliminary steps in the chapter on exhaust system design in the next revision of the *Industrial Ventilation Manual*, emphasizing the need to evaluate potential incompatibilities between dusts, fumes, or vapors that are likely to be intermixed in main header ducts of a ventilation system to ensure that they do not result in fire or explosion hazards, or destructive corrosion. Reference appropriate methods for such evaluations, such as the ASTM E 2012-00 standard. (2003-08-I-RI-R7)

By the
U.S. Chemical Safety and Hazard Investigation Board

Carolyn W. Merritt
Chair

John S. Bresland
Member

Rixio Medina
Member

Gerald V. Poje, Ph.D.
Member

Gary Lee Visscher
Member

August 20, 2004

10.0 References

American Conference of Governmental Industrial Hygienists (ACGIH), 2001. *Industrial Ventilation: A Manual of Recommended Practice*, Sections 5.1–5.2.

American National Standards Institute (ANSI), 1991. *Design and Operation of Local Exhaust Ventilation Systems*, ANSI Z9.2.

American Society for Testing and Materials (ASTM), 2003. *Standard Guide for the Preparation of a Binary Chemical Compatibility Chart*, ASTM E 2012-00.

Building Officials and Code Administrators International, Inc. (BOCA), 1987. *National Mechanical Code.*

Bodurtha, F. T., 1976. "Explosion Hazards in Pollution Control," *Loss Prevention*, Vol. 10, American Institute of Chemical Engineers (AIChE).

Bretherick, Leslie, P. G. Urben, and Martin J. Pitt, 1999. *Bretherick's Handbook of Reactive Chemical Hazards,* Sixth Edition, Vol. 1, Butterworth-Heinemann.

Center for Chemical Process Safety (CCPS), 1995. *Guidelines for Technical Planning for On Site Emergencies*, AIChE.

CCPS, 1993, *Guidelines for Engineering Design for Process Safety*, AIChE.

CCPS, 1989. *Technical Management of Chemical Process Safety,* AIChE.

Center for Waste Reduction Technology (CWRT), 2003. *Practical Solutions for Reducing Volatile Organic Compounds and Hazardous Air Pollutants*, AIChE.

Clark, D. G., and R. W. Syvester, 1996. "Ensure Process Vent Collection System Safety," *Chemical Engineering Progress*, January 1996, pp. 65–77.

Ennis, John L., and, Edward S. Shanley, 1991. "On Hazardous Silver Compounds," *Journal of Chemical Education*, Vol. 68, January 1991, p. A6.

Factory Mutual Insurance Company (FM Global), 2000. *Industrial Exhaust Systems*, Property Loss Prevention Data Sheet 7-78.

Federal Emergency Management Agency and U.S. Fire Administration (FEMA–USFA), 2003. *Risk Management Planning for Hazardous Materials: What It Means for Fire Service Planning*, USFA-TR-124, January 2003.

FEMA, 2002. *Planning Protective Action Decision-Making: Evacuate or Shelter-in-Place?* prepared by Oak Ridge National Laboratory, ORNL/TM-2002/144, June 2002.

International Code Council (ICC), 2000. *International Mechanical Code*, Section 510.5.

International Program on Chemical Safety (IPCS), 1993. *International Chemical Safety Card, Silver Nitrate*, www.cdc.gov/niosh/ipcsneng/neng1116.

National Fire Protection Association (NFPA), 2000. Prevention of Fire and Dust Explosions From the Manufacturing, Procesing, and Handling of Combustible Particulate Solids, NFPA 654.

NFPA, 1999. *Standard for Exhaust Systems for Air Conveying of Vapors, Gases, Mists, and Noncombustible Particulate Solids*, NFPA 91.

NFPA, 1998. *Liquid and Solid Oxidizers*, Vol. 7, Standard 430.

Nichols, F. P., 1998. "Design of Vent Collection and Destruction Systems," *International Symposium on Runaway Reactions, Pressure Relief Design, and Effluent Handling*, G. A. Melhem and H. G. Fisher, eds., AIChE.

Ogle, R. A., A. R. Carpenter, and D. Morrison, 2004. "Lessons Learned From Fires and Explosions Involving Air Pollution Control Systems," *38th Annual Loss Prevention Symposium*, AIChE.

Ozog, H., and W. J. Enry, 2000. "Safety Hazards Associated with Air-Emission Controls," *Process Safety Progress*, Vol 19, Spring 2000, pp. 25–31.

Renner, H., 1993. "Silver, Silver Compounds, and Silver Alloys," *Ullman's Encyclopedia of Industrial Chemistry*, Vol. A24, VCH Publishers, p. 138.

Shanley, Edward S., and John L. Ennis, 1991. "The Chemistry and Free Energy of Formation of Silver Nitride," *I&EC Research*, Vol. 30, pp. 2503–2506.

U.S. Chemical Safety and Hazard Investigation Board (USCSB), 2002. *Hazard Investigation, Improving Reactive Hazard Management*, Report No. 2001-01-H, December 2002.

U.S. Environmental Protection Agency (USEPA), 1998. *Air Pollution Operating Permit Program Update*, No. 451-K-98-002, February 1998.

APPENDIX A: Technic Production and Processes

Appendix A describes the processes connected to the vent collection system and associated chemicals, and their status immediately prior to the explosion.

A.1   Silver Nitrate

Silver nitrate is produced using the Kestner unit (an enclosed vessel) and three jacketed batch reactors located on mezzanine 1. The vent collection system transports entrained silver nitrate, nitrogen oxides (NO_x), nitric acid vapor, and water vapor.

- Kestner unit: Interviews with employees revealed that during occasional process upsets, fumes from this distillation vessel overwhelmed the vent collection system and vented into the plant. In-house maintenance personnel connected the Kestner unit to the vent collection system as shown in Figure A-1. This unit was in full operation on the day of the explosion.

- Mezzanine 1: Employees stated that on occasion dark red smoke overwhelmed the vent collection system and escaped through the loose-fitting tank access door into the plant (Figure A-2). On the day of the explosion, three batches of silver nitrate were being held on mezzanine 1 for future use, but none were being processed.

Figure A-1. Kestner unit exhaust ventilation.

Figure A-2. Silver nitrate reactor on mezzanine 1 with open tank door
(following explosion and fire).

A.2   Silver Cyanide/Potassium Silver Cyanide

Silver cyanide is produced in a reactor located on the ground level between mezzanines 1 and 2. This process is manually operated without any process alarms, interlocks, or automatic shutoffs. It produces very little gas or vapor because it is not heated and does not produce heat during the reaction; however, it is connected to the vent collection system. No silver cyanide was being produced on the day of the incident.

Potassium silver cyanide is produced in three batch reactors on mezzanine 2. The solution used in this process is reused approximately 15 times. It is prepared for reuse by heating and evaporating in jacketed reactors. Evaporation concentrates the silver cyanide and reduces the batch volume so that more reactants can be added. As the solution nears the end of its usefulness, it starts to exhibit an ammonia odor that signals the need to discard it and mix up a new batch. Discarded solutions are taken to the metals reclaim areas for silver reclamation and treatment.

This process generates silver cyanide, water vapor, potassium cyanide vapor, and ammonia vapor—all of which are extracted by the vent collection system. No batches of potassium silver cyanide were being processed or evaporated on the day of the explosion.

A.3   Plating Solution Mixtures (Kettle Room)

The kettle room is used to process made-to-order batches of some 1,700 different plating-related chemicals. It contains 20 jacketed process kettles, ranging from 60 to 1,000 gallons; and four product mixing and storage tanks, ranging from 500 to 1,000 gallons. Each of the kettles is used for a variety of batch recipes. All of the kettles and mixing tanks are connected to the vent collection system.

The most common chemicals associated with the most frequently produced batches include ammonium citrate, sulfonates, ethers, aldehydes, aliphatics, amines, alcohols, and various salts and acids. On the day of the incident, three small batches of plating chemicals were being prepared in the kettle room.

A.4   Other Connected Process Areas

The following process areas are located in the older portion of the facility and were connected to the vent collection system during the 1990 expansion.

A.4.1   Metals Reclaim Areas

Technic recovers precious metals from various waste materials in two metals reclaim areas. These wastes include spent solutions and discarded electronic components from other Technic plants. In addition, Technic receives spent plating solutions that are classified and shipped as hazardous waste under EPA and U.S. Department of Transportation (DOT) guidelines. All tanks are equipped with exhaust hoods directly connected to the vent collection system.

A.4.1.1   Silver Electroplating

During the silver electroplating[1] process, solutions are chemically treated with a formaldehyde solution to remove residual cyanide.[2] Once the formaldehyde is added, the solution is heated to approximately 180 degrees Fahrenheit (°F). The electrodes are then energized and the solution sits undisturbed overnight with the ventilation system running. Each morning, at the beginning of the shift, these solutions are pumped back into drums for disposal, and the silver is harvested from the copper cathodes. Technic performs silver electroplating daily in four to six tanks.

---

[1] At Technic, electroplating is accomplished by passing a direct electrical current through the spent solution; the dissolved silver attaches to copper plates (cathodes) submerged in the solution.

[2] Para-formaldehyde prills are a powdered form of formaldehyde added to cyanide-containing solutions to reduce the threat of generating hydrogen cyanide gas (which is extremely toxic) and to make the solutions less hazardous for disposal.

A.4.1.2   Gold Precipitating

Waste gold-plating solutions are chemically treated to prevent toxic gas generation and to precipitate the dissolved gold out of solution.[3] Once pretreated, sodium hydrosulfite is added to the solution to activate the precipitation process. The solution is then allowed to sit overnight. At the beginning of the next shift, the reacted cyanide is skimmed off and the remaining solution is pumped into barrels for disposal. The gold is then pumped from the bottom of the tank.

A high pH cyanide solution is used to strip the gold from plated electronic circuitry. This process creates potassium gold cyanide, which is then put through the gold precipitation process (described above) to recover the dissolved gold.

A.4.2   Cyanide Salt De-lumping and Packaging (Weigh Room)

Silver cyanide and potassium silver cyanide pastes produced on mezzanines 1 and 2 are dried in two ovens in this area.[4] After drying to the consistency of caked powder, the materials are de-lumped in grinders and then poured into 55-gallon drums equipped with auger-type dispensers for packaging. The exhausts from the grinders pass through filters that exhaust to the vent collection system. The exhaust vents from the dispensers pass through an in-line baghouse (removed in December 2002) connected to the vent collection system.

---

[3] Potassium hydroxide flakes are added to stabilize the pH of the solution and to prevent the generation of hydrogen cyanide and chlorine gases.

[4] Oven exhausts are not connected to the ventilation system.

APPENDIX B: Calorimetric Testing

Under contract to CSB, ioMosaic Corporation, Salem, New Hampshire, conducted three accelerating rate calorimetry (ARC) tests to explore the stability and chemical reaction characteristics of silver nitrate with other organics that were present in the facility in large quantities and could have been present in the vent collection system:

- Sodium hydrosulfite/silver nitrate, equimolar mixture.

- Ammonium citrate/silver nitrate, equimolar mixture.

- Carbon/silver nitrate, 1.5:1 molar ratio.

The ARC was selected as a test instrument because of its ability to detect exothermic activity at low self-heating rates in a near-adiabatic test environment.

The ARC test data indicate that the sodium hydrosulfite/silver nitrate was the most sensitive mixture. Small levels of self-heating were detected at 35 degrees Celsius (°C), as evident by the ARC trace. Mixture 3 had the highest onset temperature (170°C; less sensitive than mixture 1), but once initiated led to an actual explosion that destroyed the test cell. Mixture 2 onset was detected at 130°C. The data clearly show that a violent reaction between silver nitrate and other organics can be initiated at temperatures as low as 35°C. Although these tests were conducted in a near adiabatic environment, the results are directly applicable to the Technic vent collection system because the reacting materials could have been present in the system under inertial confinement.

Comparison of silver nitrate reactions with carbon, sodium hydrosulfite, and ammoniun citrate

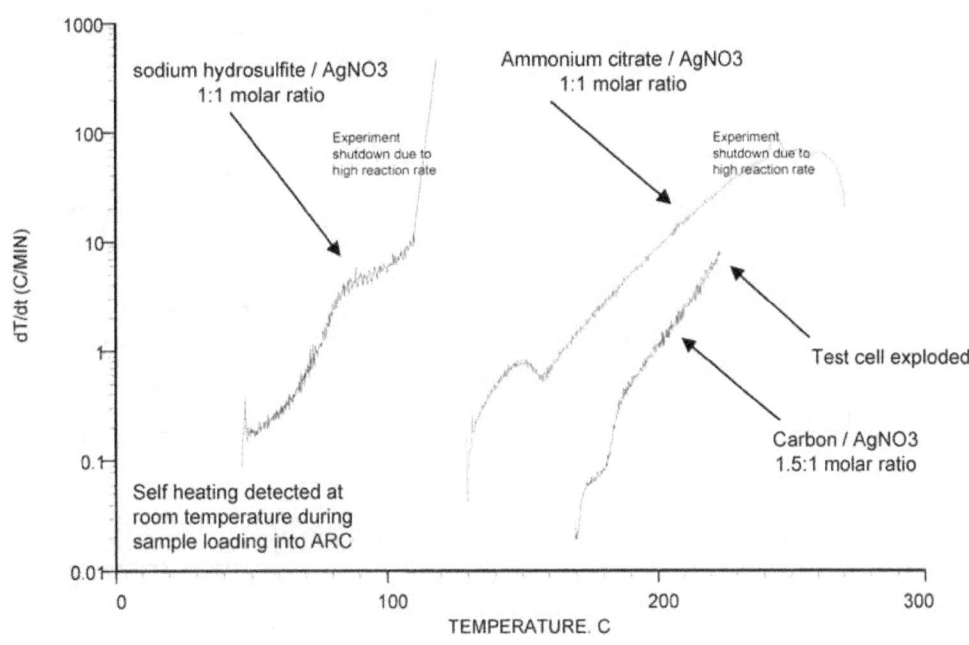

Comparison of silver nitrate reactions with carbon, sodium hydrosulfite, and ammoniun citrate

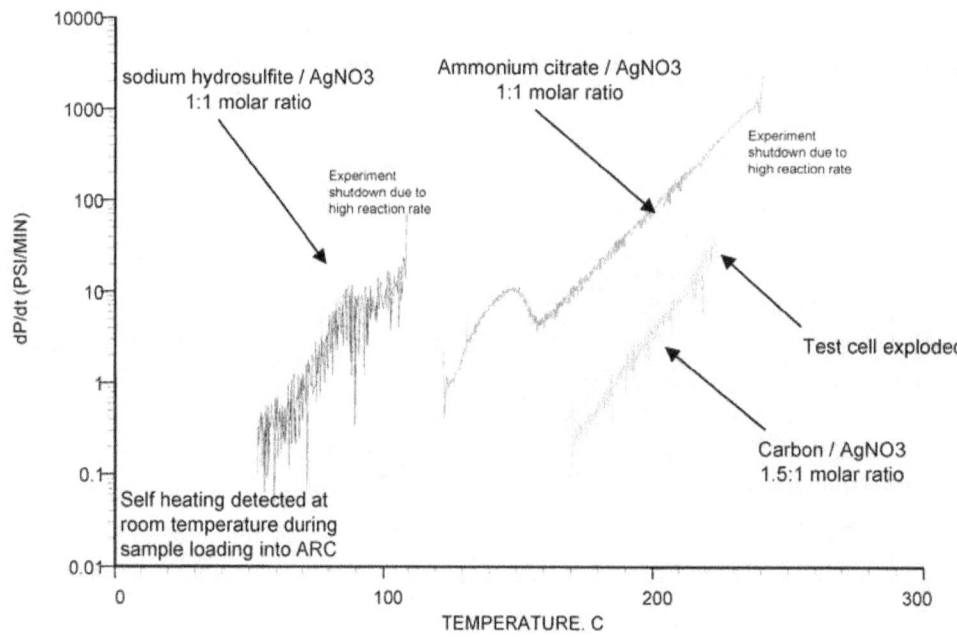

RVC/XRD Analysis of Explosion Incident/25-Jun-03   9

APPENDIX C: Causal Factors Diagram

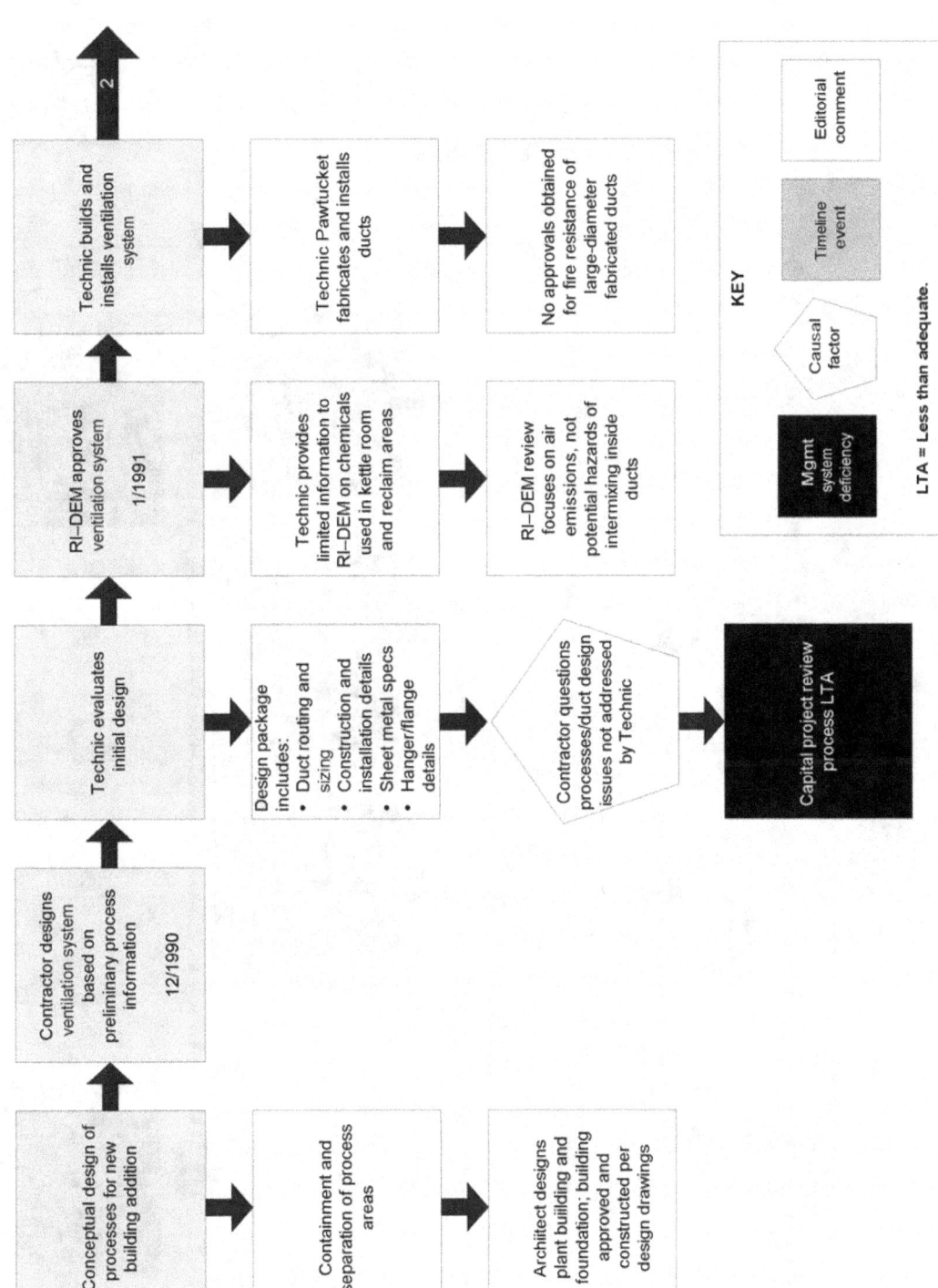

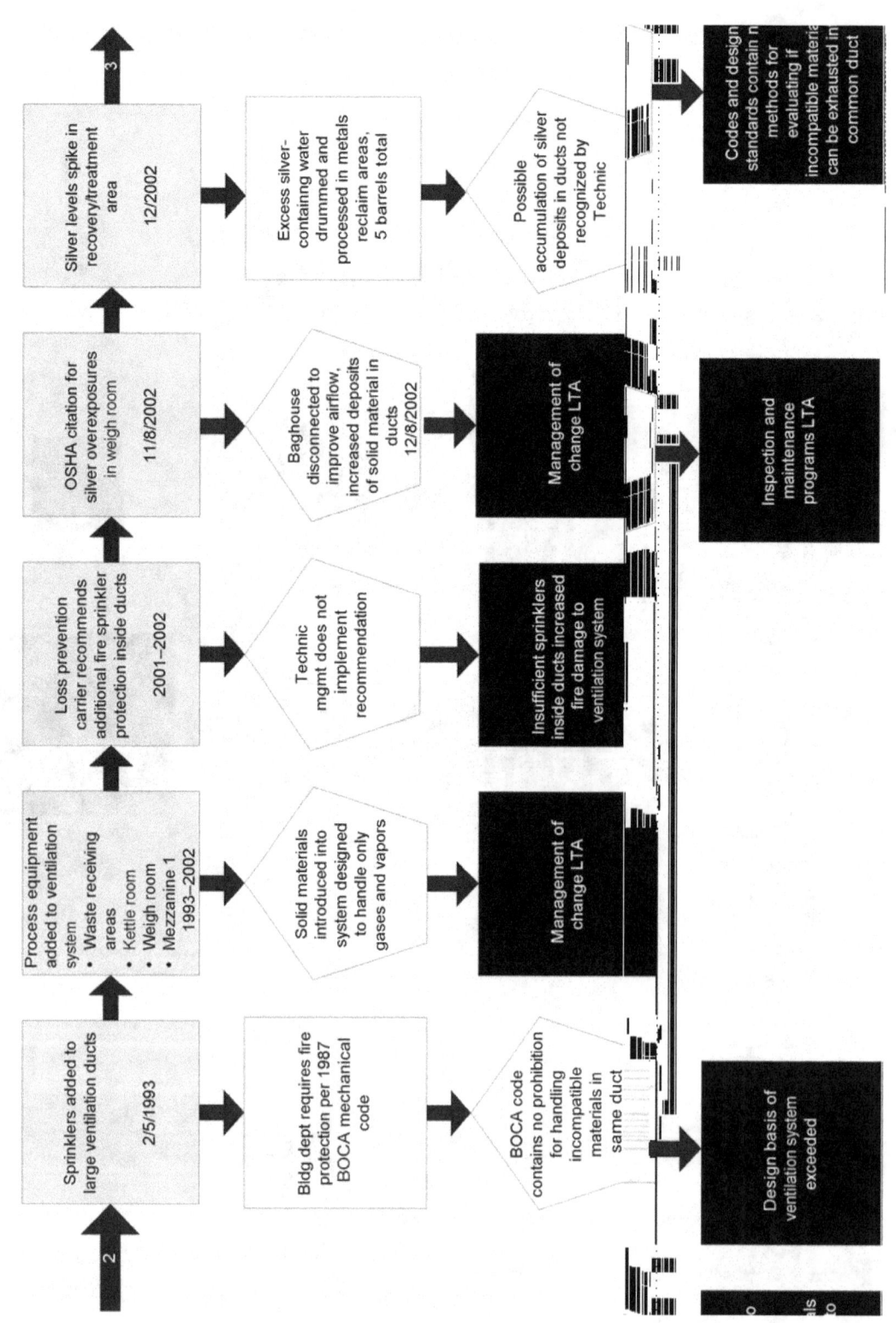

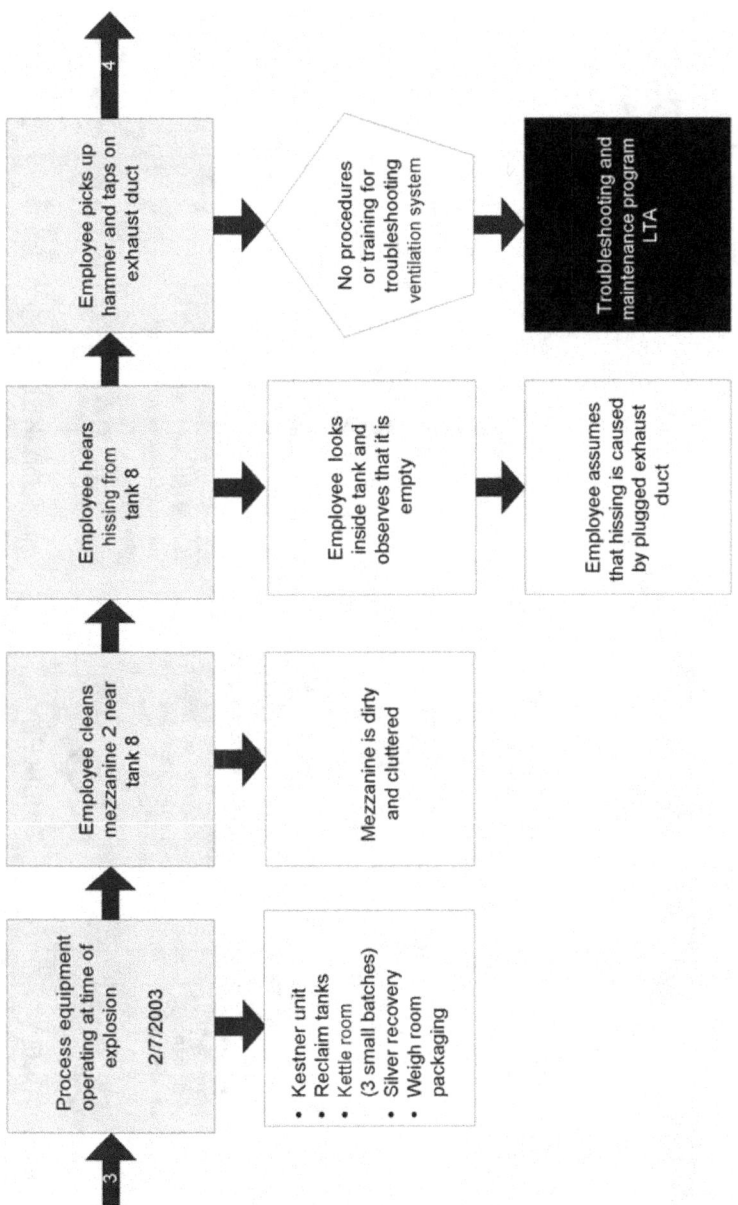

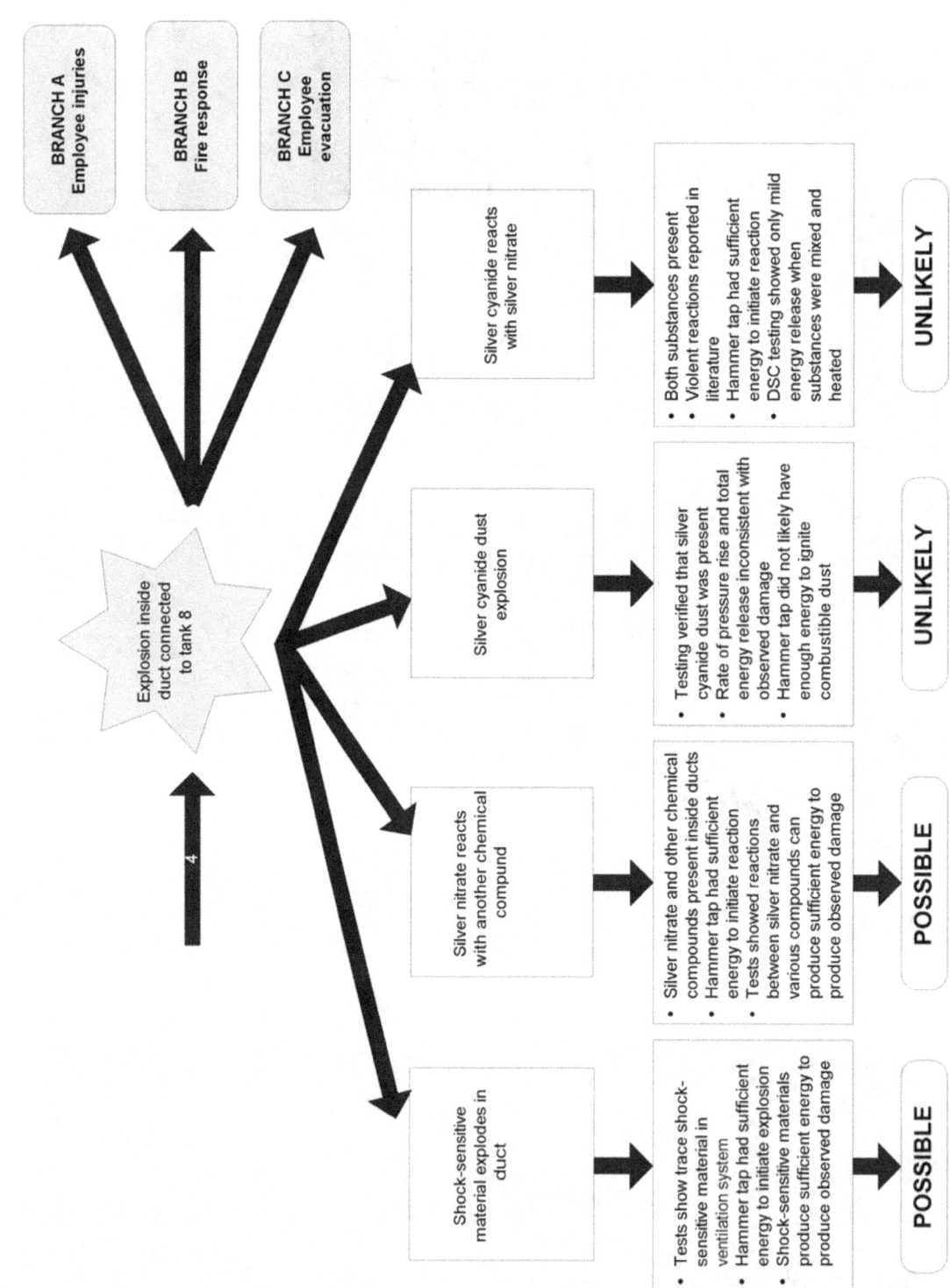

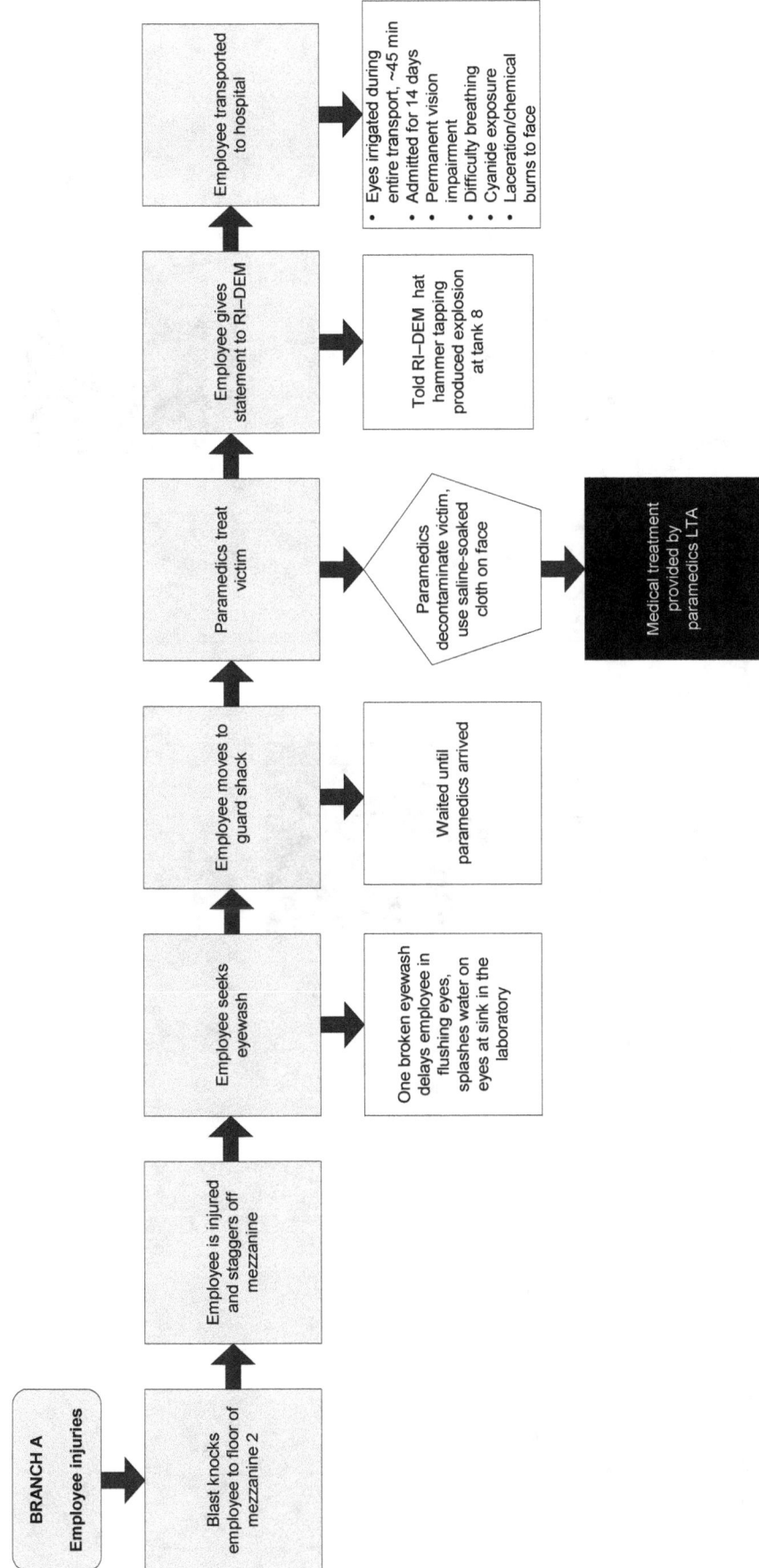

**BRANCH A
Employee injuries**

Blast knocks employee to floor of mezzanine 2

Employee is injured and staggers off mezzanine

Employee seeks eyewash
- One broken eyewash delays employee in flushing eyes, splashes water on eyes at sink in the laboratory

Employee moves to guard shack
- Waited until paramedics arrived

Paramedics treat victim
- Paramedics decontaminate victim, use saline-soaked cloth on face
- Medical treatment provided by paramedics LTA

Employee gives statement to RI–DEM
- Told RI–DEM hat hammer tapping produced explosion at tank 8

Employee transported to hospital
- Eyes irrigated during entire transport, ~45 min
- Admitted for 14 days
- Permanent vision impairment
- Difficulty breathing
- Cyanide exposure
- Laceration/chemical burns to face

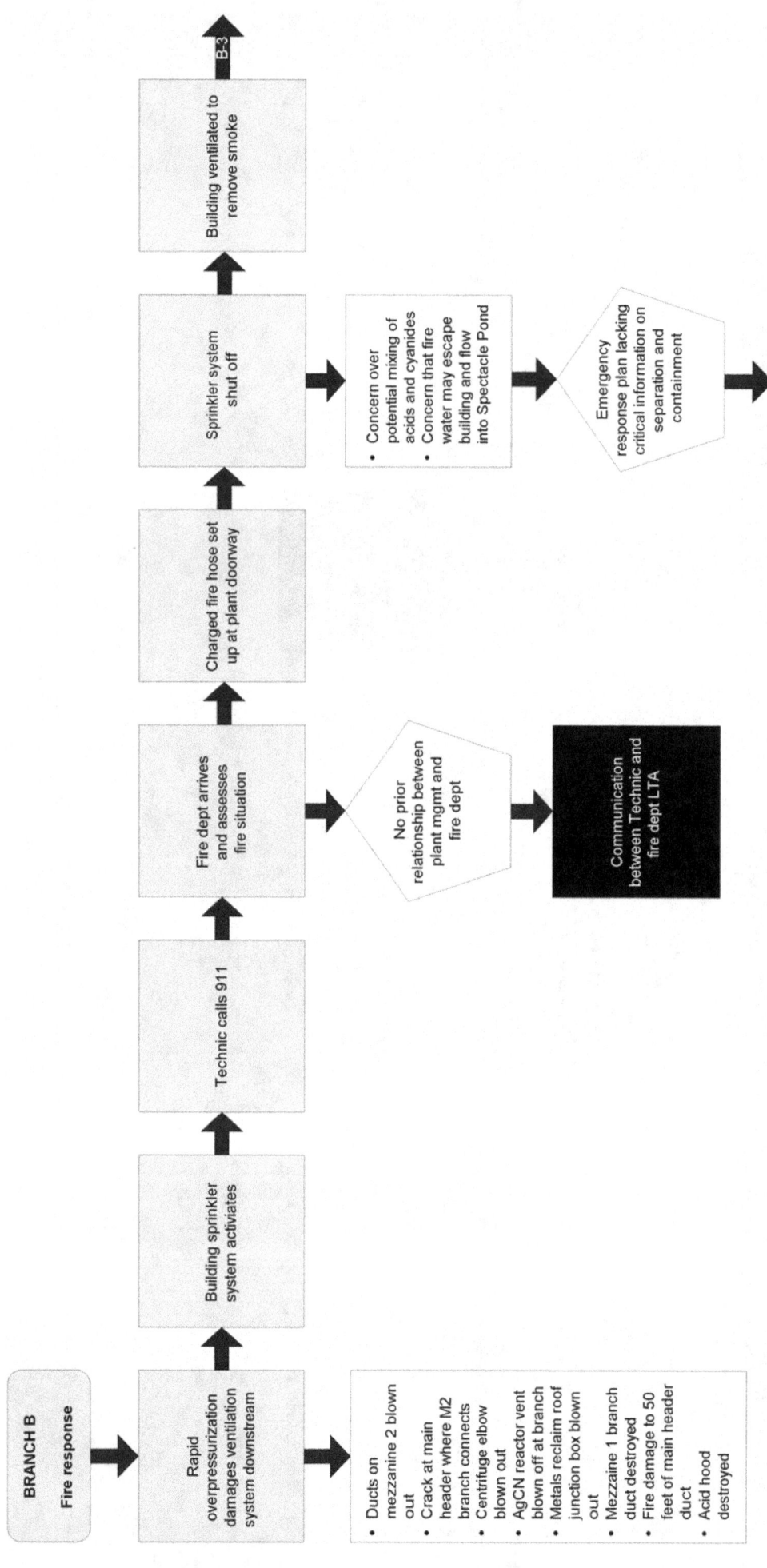

BRANCH B
Fire response

Rapid overpressurization damages ventilation system downstream

- Ducts on mezzanine 2 blown out
- Crack at main header where M2 branch connects
- Centrifuge elbow blown out
- AgCN reactor vent blown off at branch
- Metals reclaim roof junction box blown out
- Mezzaine 1 branch duct destroyed
- Fire damage to 50 feet of main header duct
- Acid hood destroyed

Building sprinkler system activiates

Technic calls 911

Fire dept arrives and assesses fire situation

No prior relationship between plant mgmt and fire dept

Communication between Technic and fire dept LTA

Charged fire hose set up at plant doorway

Sprinkler system shut off

- Concern over potential mixing of acids and cyanides
- Concern that fire water may escape building and flow into Spectacle Pond

Emergency response plan lacking critical information on separation and containment

Technic emergency response plan LTA

Building ventilated to remove smoke

B3

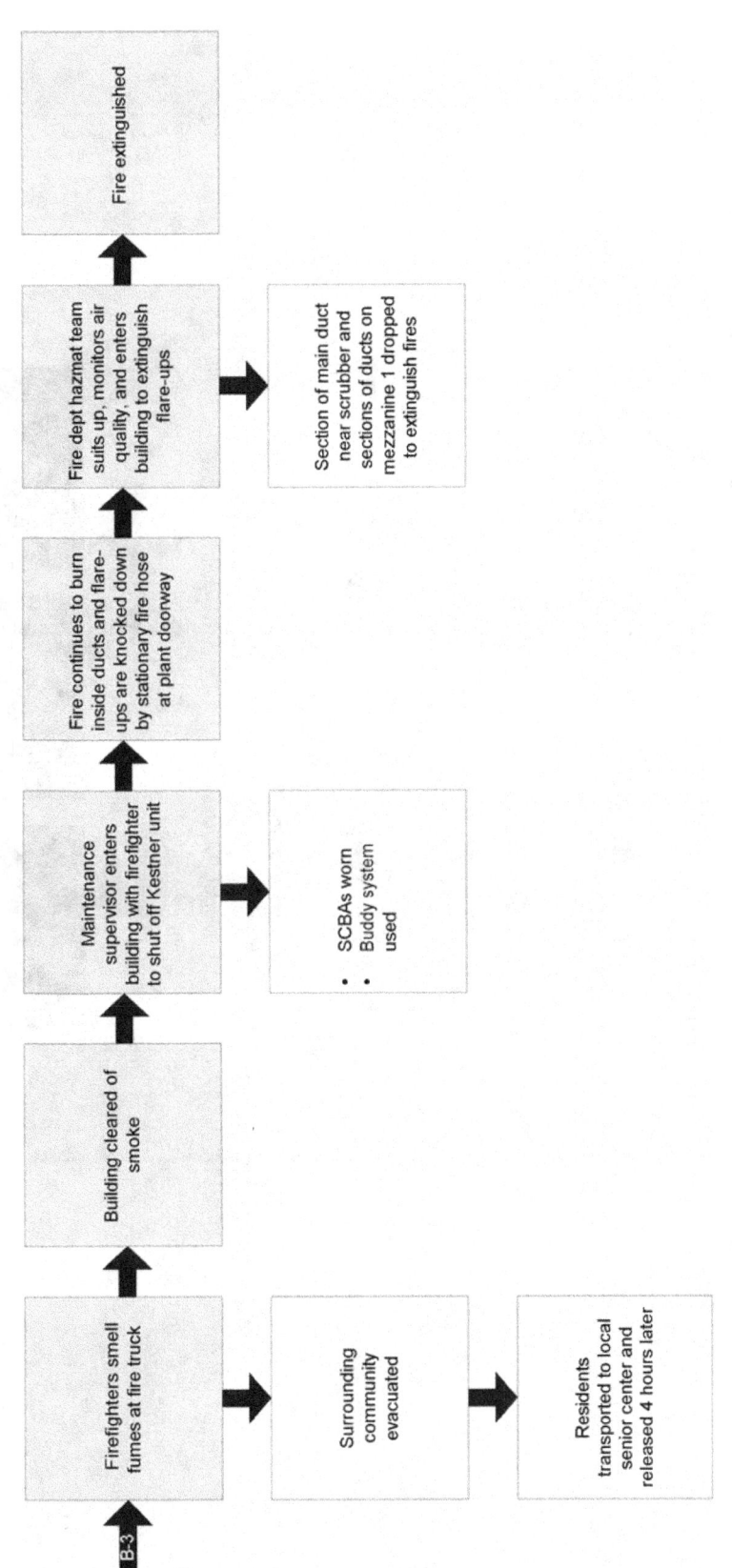

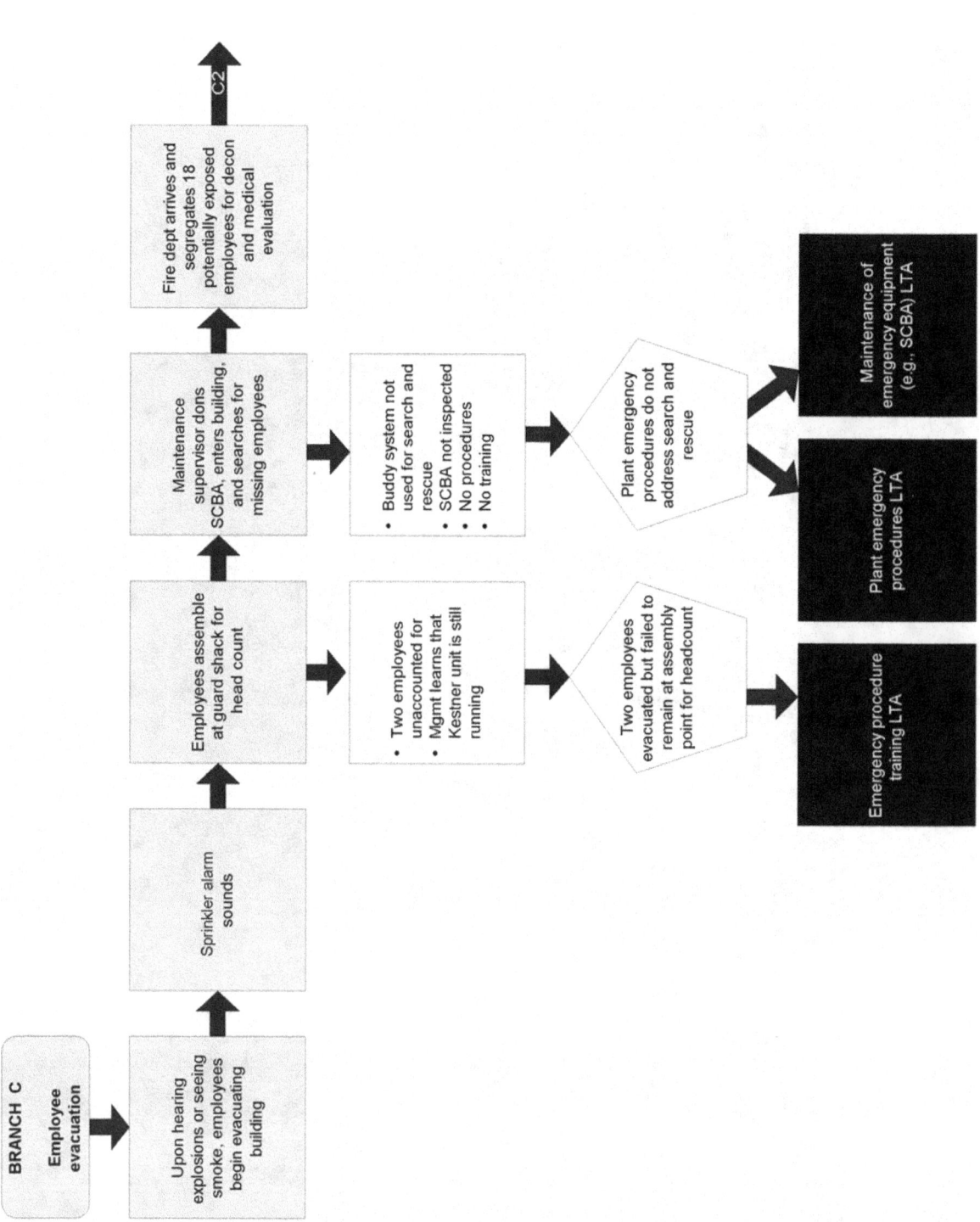

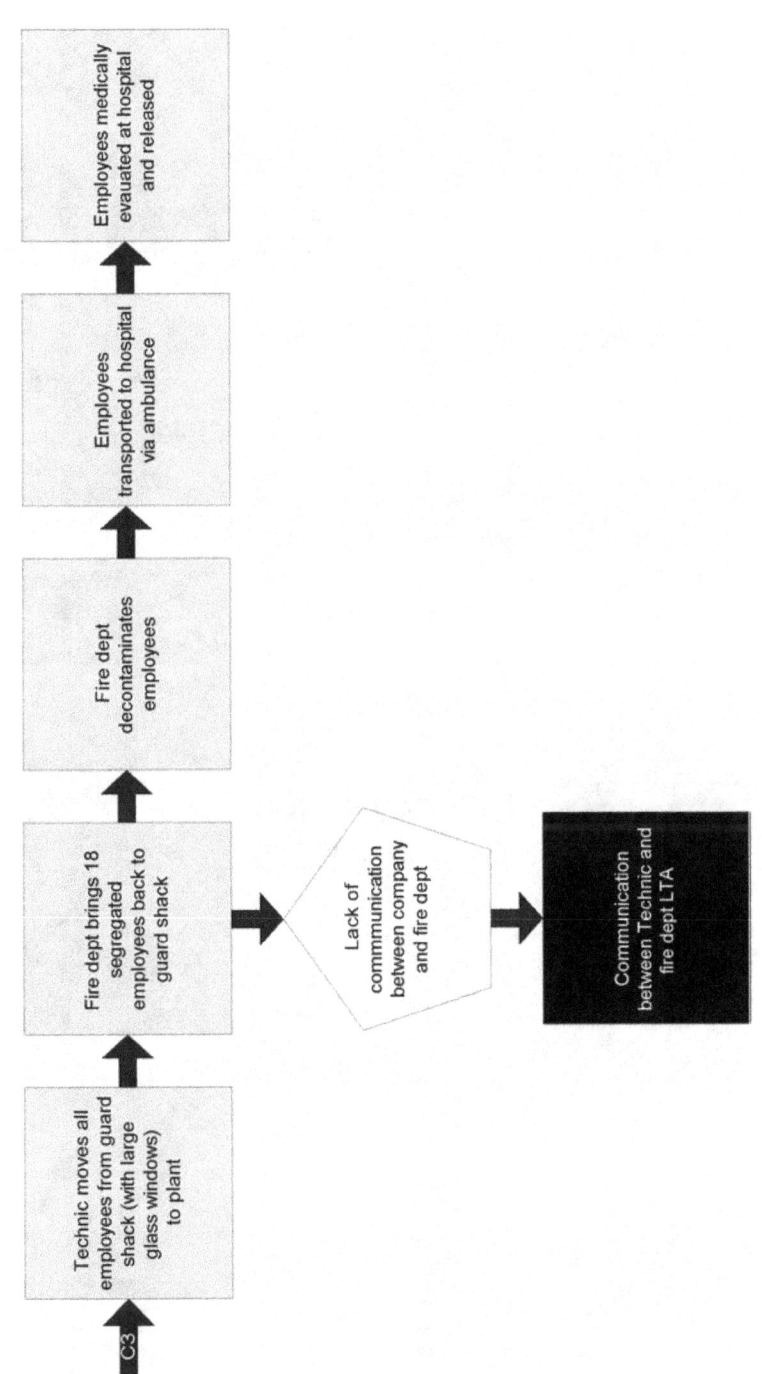